Test Item

by Rex Joyner

PHYSICS

Principles with Applications

FOURTH EDITION

Douglas C. Giancoli

Prentice Hall Englewood Cliffs, NJ 07632

© 1995 by **PRENTICE-HALL, INC.**
A Simon & Schuster Company
Englewood Cliffs, N.J. 07632

10 9 8 7 6 5 4 3 2 1

ISBN 0-13-141301-5

Printed in the United States of America

CONTENTS

PRENTICE HALL: INSTRUCTOR SUPPORT FOR TEST ITEM FILES

This hard copy test item file is just one part of Prentice Hall's comprehensive testing support service, which also includes:

1. *Prentice Hall Test Manager 2.0:* This powerful computerized testing package is available for DOS-based computers in either 3.5 or 5.25 format. It offers full mouse support, complete question editing, random test generation, graphics and printing capabilities. Toll free technical support is offered to all users, and the Test Manager is free. You may contact your local rep or call our Faculty Support Services department at 1-800-333-7945. Please identify the main text author, title, and disk size. *Some* test item files are also available on Macintosh.

2. For those instructors without access to a computer, we offer the popular *Prentice Hall Telephone Testing Service:* It's simple, fast, and efficient. Simply pick the questions you'd like on your test from this bank, and call our College Media department at 1-800-842-2958, outside the U.S. call 1-201-592-3263. Identify the main text and test questions you'd like, as well as any special instructions. We will create the test (or multiple versions if you wish) and send you a master copy for duplication within 48 hours. Free to adopters for life of text use.

. Introduction

1.1: Match the physical unit to the the physical quantity.

1.1: kilogram

Answer: 1000 g
MA Easy

1.2: nano-second

Answer: 10^{-9} s
MA Easy

1.3: conversion

Answer: 1 in. = 2.54 cm
MA Easy

1.4: scientific notation

Answer: 0.062 m = 6.2 X 10^{-2} m
MA Easy

1.5: mega-watts

Answer: 10^6 W
MA Easy

1.2: Substitute the correct prefix to the fun-word.

1.6: 10^{-3} -mouse

Answer: milli
MA Easy

1.7: 10 -cards

Answer: deka
MA Easy

1.8: 10^{15} -pan

Answer: peta
MA Easy

1

1.9: 10^{-18} -money

Answer: atto
    MA   Easy

1.10: 10^3 -mockingbird

Answer: kilo
    MA   Easy

1.11: 10^{-1} and Lucy

Answer: deci
    MA   Easy

1.12: 10^{-12} -low

Answer: pico
    MA   Easy

1.13: 10^{-21}, Harpo, Chico, & Groucho

Answer: zepto
    MA   Easy

1.14: 10^9 -ly

Answer: giga
    MA   Easy

1.15: 10^{12} -firma

Answer: tera
    MA   Easy

1.16: 10^{24} -see me now

Answer: yotta
    MA   Easy

1.17: $10^{18}2$ -terrestrial

Answer: exa
    MA   Easy

1.18: 10^{15}, Paul, & Mary

Answer: peta
    MA   Easy

2

1.19: 10^{18} -lent

Answer: exa
        MA   Easy

1.20: 10^{12} -tory

Answer: tera
        MA   Easy

1.21: All of the following are base units of the SI system except:
      a) kilogram.   b) kelvin.   c) meter.   d) volt.

Answer: (d)
        MC   Easy

1.22: 100 mL is equal to
      a) 1 kL   b) $10^{-6} \mu L$ c) 0.1 L   d) 0.01 ML

Answer: (c)
        MC   Easy

1.23: The SI prefix for 10^{-12} is
      a) tera-   b) giga-   c) nano-   d) pico-

Answer: (d)
        MC   Easy

1.24: Given (1 angstrom unit = 10^{-10} m) and (1 fermi = 10^{-15} m), what
      is the relationship between these units?
      a) 1 angstrom = 10^5 fermi   b) 1 angstrom = 10^{-5} fermi
      c) 1 angstrom = 10^{-25} fermi   d) 1 angstrom = 10^{+25} fermi

Answer: (a)
        MC   Easy

1.25: A football field is 120 yd long and 50 yd wide. What is the area
      of the football field, in m^2, if 1 yd = 91.44 cm?
      a) 2400 m^2 b) 3688 m^2 c) 4206 m^2 d) 5019 m^2

Answer: (d)
        MC   Moderate

1.26: An average human has a heart rate of 70 beats per minute. If
      someone's heart beat at that average rate over a 70-yr lifetime,
      how many times would it beat?
      a) 7.44 x 10^5 b) 2.20 x 10^6 c) 1.78 x 10^7 d) 2.58 x 10^9

Answer: (d)
        MC   Moderate

1.27: The last page of a book is numbered 764. The book is 3 cm thick. What is the average thickness of a sheet of paper in the book, in centimeters?
   a) 0.0039   b) 0.0078   c) 127.3   d) 254.7

Answer: (b)
       MC   Moderate

1.28: A sunken treasure sits on the ocean floor at a depth of 600 fathoms. What is this depth, in feet? (1 fathom = 6 ft)
   a) 100 ft   b) 600 ft   c) 1200 ft   d) 3600 ft

Answer: (d)
       MC   Easy

1.29: An astronomical unit (AU) is equal to the average distance from the earth to the sun, about 92.9×10^6 mi. A parsec (pc) is the distance at which 1 AU would subtend an angle of 1 second of arc. (a) How many miles are there in a parsec? (b) How many AUs are there in a parsec?

Answer: (a) 1.92×10^{13} miles = 1 parsec
       (b) 2.07×10^5 AUs = 1 parsec
       ES   Moderate

1.30: Which of the following has three significant figures?
   a) 305.0 cm   b) 0.0500 mm
   c) 1.00081 kg   d) 8.060×10^{11} m^2

Answer: (b)
       MC   Easy

1.31: The number of significant figures in 0.40 is
   a) one.   b) two.   c) three.   d) four.

Answer: (b)
       MC   Easy

1.32: Determine the number of significant figures in each of the following measured numbers:
   (a) 4.61 cm   (b) 24.0 s   (c) 0.055 ms   (d) 100.01 m

Answer: (a)   3   (b) 3   (c) 2   (d) 5
       ES   Easy

1.33: Express each of the following numbers to three significant figures: (a) 21.22 m   (b) 208.7 kg   (c) 0.0015601 g   (d) 221 s

Answer: (a)   21.2 m   (b)   209 kg   (c) 0.00156 g   (d) 221 s
       ES   Easy

1.34: 0.0097×10^6 cm equals 97×10^{10} cm.

Answer: False
       TF   Easy

1.35: 0.00325×10^{-8} cm equals
       a) 3.25×10^{-12} mm   b) 3.25×10^{-11} mm
       c) 3.25×10^{-10} mm   d) 3.25×10^{-9} mm

Answer: (c)
       MC   Moderate

1.36: The number of significant figures a common ruler can measure is
       a) one.   b) three.   c) five.   d) seven.

Answer: (b)
       MC   Moderate

1.37: Four students measure the mass of an object, each using different
       scale. They record their results as follows:

student	A	B	C	D
mass (g)	49.06	49	50	49.2

       Which student used the least precise scale?
       a) A   b) B   c) C   d) D

Answer: (c)
       MC   Moderate

1.38: Express the result of the following calculation, to the proper
       number of significant figures:
       $(0.02739) \times (-240{,}000) =$

Answer: -6,600
       ES   Easy

1.39: Express the result of the following calculation, in scientific
       notation, to the proper number of significant figures:
       $((395600.1)/(6.72)) =$

Answer: 5.89×10^4
       ES   Easy

1.40: Wall posters are usually sold curled up in cylindrical cardboard
       tubes. If the length of the tube is 84.5 cm, and the diameter of
       the tube is 2.40 cm, what is the area of the poster, in cm^2?
       (Assume the poster doesn't overlap itself.)
       a) 203 cm^2 b) 382 cm^2 c) 637 cm^2 d) 1529 cm^2

Answer: (c)
       MC   Moderate

1.41: 0.0001993 is the same as
   a) 1.993×10^{-4}
   b) 19.93×10^5
   c) 1993×10^7
   d) 199.3×10^2

Answer: (a)
   MC   Easy

1.42: The average density of blood is 1.06×10^3 kg/m^3. If you donate a pint of blood to the Red Cross, what mass of blood have you donated, in grams?
   (1 pt = 1/2 L, 1 L = 1000 cm^3)
   a) 530 g
   b) 0.530 g
   c) 5.30×10^3 g
   d) 5.30×10^5 g

Answer: (a)
   MC   Moderate

1.43: The mass of Mars, 6.40×10^{23} kg, is about one-tenth that of the earth, and its radius, 3395 km, is about half that of earth. What is the mean density of Mars?

Answer: 3900 kg/m^3
   ES   Moderate

1.44: Concrete is sold by the cubic yard. What is the mass, in kilograms, of one cubic yard of concrete that is five times as dense as water? (1 m = 1.094 yd, and 1 m^3 of water has a mass of 1,000 kg.)
   a) 764 kg   b) 2420 kg   c) 3819 kg   d) 6546 kg

Answer: (c)
   MC   Moderate

1.45: A thick-walled metal pipe of length 20 cm has an inside diameter of 2.0 cm and an outside diameter of 2.4 cm. What is the total surface area of the pipe, counting the ends?

Answer: 279 cm^2
   ES   Moderate

1.46: The metric prefix for one thousand is
   a) milli   b) centi   c) kilo   d) mega

Answer: (c)
   MC   Easy

1.47: Fifty milligrams is
   a) 0.00005 g   b) .0005 g   c) .05 g   d) 50000 g

Answer: (c)
   MC   Easy

1.48: Forty kilometers is
a) 400,000 m   b) 40,000 m   c) 40 m   d) 0.04 m

Answer: (b)
MC   Easy

1.49: The metric prefix for one million is milli-.

Answer: False
TF   Easy

1.50: 90 milliseconds is 0.090 s.

Answer: True
TF   Easy

1.51: The metric prefix for one million is
a) mega-   b) milli-   c) micro-   d) nano-

Answer: (a)
MC   Easy

1.52: The metric prefix deca- means
a) 1000   b) 100   c) 10   d) 0.1

Answer: (c)
MC   Easy

1.53: In the MKS system, the following are fundamental units:
a) Newton, meter, second   b) gram, centimeter, second
c) kilogram, meter, hour   d) kilogram, meter, second

Answer: (d)
MC   Moderate

1.54: The metric prefix for one hundred is
a) hecto-   b) milli-   c) micro-   d) kilo-

Answer: (a)
MC   Easy

1.55: The metric prefix for one one-thousandth is
a) hecto-   b) milli-   c) micro-   d) kilo-

Answer: (b)
MC   Easy

1.56: What does the metric prefix deci- mean?
   a) 100   b) 10   c) 0.1   d) 0.01

Answer: (c)
   MC   Easy

1.57: How many centimeters are in a kilometer?
   a) 1,000,000   b) 100,000   c) 1000   d) 100

Answer: (b)
   MC   Moderate

1.58: You measure the length and width of a rectangle as 1.125 m and
   0.606 m, respectively. Multiplying, your calculator gives the
   product as 0.68175. Rounding properly to the correct number of
   significant figures, the area should be written as
   a) 0.68 m^2 b) 0.682 m^2 c) 0.6818 m^2 d) 0.68175 m^2

Answer: (b)
   MC   Easy

1.59: You measure the length and width of a rectangle to be 1.125 m and
   0.606 m, respectively. You calculate the rectangle's perimeter
   by adding these and multiplying by two. Your calculator's
   display reads 3.462. To the correct number of significant
   figures, this should be written as
   a) 3.5 m   b) 3.46 m   c) 3.462 m   d) 3.4620 m

Answer: (c)
   MC   Easy

1.60: If 1 inch = 2.54 cm exactly, how many square centimeters are in
   1.00000 square feet?
   a) 929.030 cm^2 b) 365.760 cm^2
   c) 30.4800 cm^2 d) 6.45160 cm^2

Answer: (a)
   MC   Moderate

1.61: How many basic units does the SI system have?
   a) four   b) five   c) seven   d) ten

Answer: (c)
   MC   Easy

1.62: Select the list which contains only SI basic units.
   a) liter, meter, second, watt
   b) joule, kelvin, kilogram, watt
   c) candela, kelvin, meter, second
   d) joule, newton, second, watt

Answer: (c)
   MC   Easy

1.63: A useful method of expressing very small or very large numbers is
    a) scientific notation   b) arabic numerals
    c) the metric system   d) roman numerals

Answer: (a)
    MC   Easy

1.64: 1000 is correctly expressed in scientific notation as
    a) 0.1×10^2 b) 1×10^3 c) 10×10^3 d) 1000×10^3

Answer: (b)
    MC   Easy

1.65: 4567.89 is properly expressed in scientific notation as
    a) 4.56789×10^3 b) 45.6789×10^2
    c) 456.789×10^1 d) 4567.89×10^0

Answer: (a)
    MC   Easy

1.66: 0.00048 is correctly expressed in scientific notation as
    a) 0.00048×10^4 b) 0.00048×10^{-4}
    c) 4.8×10^4 d) 4.8×10^{-4}

Answer: (d)
    MC   Easy

1.67: Convert 1.2×10^{-3} to decimal notation.
    a) 1.200   b) 0.1200   c) 0.0120   d) 0.0012

Answer: (d)
    MC   Easy

1.68: Convert 3.865×10^5 to decimal notation.
    a) 3865   b) 38650   c) 386500   d) 386500000

Answer: (c)
    MC   Easy

1.69: Select the smallest value.
    a) 15×10^{-3} b) 0.15×10^0
    c) 0.00015×10^3 d) 0.00000015×10^6

Answer: (a)
    MC   Moderate

1.70: The basic SI unit of length is the
    a) millimeter   b) centimeter   c) meter   d) kilometer

Answer: (c)
    MC   Easy

1.71: The most appropriate metric unit of length to use to measure the length of a common pencil would be the
a) micrometer   b) centimeter   c) meter   d) kilometer

Answer: (b)
MC   Easy

1.72: The most appropriate metric unit to measure the width of a city street would be the
a) micrometer   b) centimeter   c) meter   d) kilometer

Answer: (c)
MC   Easy

1.73: Convert 0.75 kilometers to meters.
a) 7500 m   b) 750 m   c) 75 m   d) 0.00075 m

Answer: (b)
MC   Easy

1.74: Convert 0.0215 m to mm.
a) 0.215 mm   b) 2.15 mm   c) 21.5 mm   d) 215 mm

Answer: (c)
MC   Easy

1.75: If 1 in = 2.54 cm, and 1 yd = 36 in, how many meters are in 7.000 yd?
a) 6.401 m   b) 36.28 m   c) 640.1 m   d) 1778 m

Answer: (a)
MC   Moderate

1.76: Which is the largest area?
a) 2,500,000 cm   b) 100,000 cm^2
c) 7500 m   d) 0.75 m^2

Answer: (b)
MC   Easy

1.77: Add 321.475, 42.500, and 2.25, and round properly.
a) 366   b) 366.2   c) 366.23   d) 366.225

Answer: (c)
MC   Moderate

1.78: Multiply 12.75 times 4.375 and round properly.
a) 55.78125   b) 55.78   c) 55.8   d) 56.0

Answer: (b)
MC   Moderate

1.79: A rectangle is 3.25 m long and 1.5 m wide. What is its area, properly rounded?
   a) 4.875 m^2 b) 4.87 m^2 c) 4.80 m^2 d) 4.9 m^2

Answer: (d)
   MC Moderate

1.1: Match the physical unit to the the physical quantity.
   TA

1.2: Substitute the correct prefix to the fun-word.
   TA

2. Describing Motion: Kinetics in One Dimension

2.1: An ly (light year) is the distance that light travels in one year. The speed of light is 3.00×10^8 m/s. How many miles are there in an ly?
(1 mi = 1609 m), (1 yr = 365 days)
a) 9.46×10^{12} mi
b) 9.46×10^{15} mi
c) 5.88×10^{12} mi
d) 5.88×10^{15} mi

Answer: (c)
MC   Moderate

2.2: How many seconds would it take light, traveling at a speed of 186,000 mi/s, to reach us from the sun? The sun is 93,000,000 mi from the earth.
a) 500 s
b) 250 s
c) 1.7×10^{13} s
d) 2 s

Answer: (a)
MC   Easy

2.3: I ran my fastest marathon (42.0 km) in 2^{hrs}:57^{min}. My average speed, in m/s, was
a) 14.2×10^3 m/s
b) 124 m/s
c) 3.95 m/s
d) 14.2 m/s

Answer: (c)
MC   Moderate

2.4: A car travels at 40 km/h for 30 min and 60 km/h for 15 min. How far does it travel in this time?
a) 20 km   b) 35 km   c) 37.5 km   d) 45 km

Answer: (b)
MC   Moderate

2.2: **Figure 2-2**

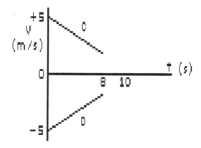

2.5: displacement

Answer: 40 km, SW
MA   Easy   Table: 2

2.6: SI acceleration of gravity

Answer: 9.8 m/s^2
        MA Easy   Table: 2

2.7: velocity

Answer: -120 m/s
        MA Easy   Table: 2

2.8: distance

Answer: 186,000 mi
        MA Easy   Table: 2

2.9: Which of the following can never be negative?
     a) Average velocity   b) Displacement
     c) Instantaneous speed   d) Acceleration of gravity

Answer: (c)
        MC Easy

2.10: If you run a complete loop around an outdoor track (400 m), in
      100 s, your average velocity is
      a) 0.25 m/s   b) 4.0 m/s   c) 40,000 m/s   d) zero

Answer: (d)
        MC Moderate

2.11: 60 mi/h equals 88 ft/s.

Answer: True
        TF Moderate

2.12: It is possible to have a zero acceleration, and still be moving.

Answer: True
        TF Easy

2.13: When is the average velocity of an object equal to the
      instantaneous velocity ?
      a) This is always true.
      b) This is never true.
      c) This is the case only when the velocity is constant.
      d) This is the case only when the velocity is increasing at a
      constant rate.

Answer: (c)
        MC Moderate

2.14: A race car circles 10 times around an 8-km track in 20 min. (a) What is its average speed per lap? (b) What is its average velocity per lap?

Answer: (a) average lap speed = 66.7 m/s   (b) average lap velocity = 0
        ES   Easy

  2.1: **Figure 2-1**

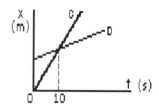

  2.15: Refer to Figure 2-1, At t = 0 s
        a) rider C is ahead of rider D.
        b) rider D is ahead of rider C.
        c) riders C and D are at the same position.

Answer: (b)
        MC   Easy   Table: 1

  2.16: Refer to Figure 2-1, At t = 0 s
        a) C is moving, and D is at rest.
        b) D is moving, and C is at rest.
        c) C and D are both moving.
        d) C and D are both at rest.

Answer: (c)
        MC   Moderate   Table: 1

  2.17: Refer to Figure 2-1, At t = 0 s
        a) C has a greater velocity than D.
        b) D has a greater velocity than C.
        c) C and D have the same velocity.
        d) C is accelerating.

Answer: (a)
        MC   Moderate   Table: 1

  2.18: Refer to Figure 2-1, At t = 10 s
        a) C and D are at the same position.
        b) C and D have the same velocity.
        c) the velocity of D is greater than the velocity of C.
        d) C is in front of D.

Answer: (a)
        MC   Easy   Table: 1

2.2: **Figure 2-2**

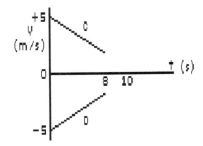

2.19: Refer to Figure 2-2, During the first 8 s
a) C has decreasing velocity, and D has increasing velocity.
b) C and D both have decreasing velocities.
c) C and D have constant velocities.
d) C has the same average velocity as D.

Answer: (a)
MC Moderate Table: 2

2.20: Refer to Figure 2-2, During the first 8 s
a) C always has a greater acceleration than D.
b) their accelerations are equal in magnitude, but opposite in sign.
c) their accelerations are equal in magnitude, and equal in sign.

Answer: (b)
MC Moderate Table: 2

2.21: Refer to Figure 2-2, Based on all the graphical information
a) they meet at the same position at t = 8 s.
b) they will meet at the same position at t = 10 s.
c) they will never meet at the same position.
d) not enough information is given to decide if they meet.

Answer: (d)
MC Moderate Table: 2

2.22: Negative velocities, approaching zero velocity, are negative accelerations.

Answer: False
TF Moderate

2.23: A new car manufacturer advertises that their car can go "from zero to sixty in 8 s". This is a description of
a) average speed.
b) instantaneous speed.
c) average acceleration.
d) instantaneous acceleration.

Answer: (c)
MC   Easy

2.24: Can an object's velocity change direction when its acceleration is constant?
a) No, this is not possible because it is always speeding up.
b) No, this is not possible because it is always speeding up or always slowing down, but it can never turn around.
c) Yes, this is possible, and a rock thrown straight up is an example.
d) Yes, this is possible, and a car that starts from rest, speeds up, slows to a stop, and then backs up is an example.

Answer: (c)
MC   Moderate

2.25: Can an object have increasing speed while its acceleration is decreasing?
a) No, this is impossible because of the way in which acceleration is defined.
b) No, because if acceleration is decreasing the object will be slowing down.
c) Yes, and an example would be an object falling in the absence of air friction.
d) Yes, and an example would be an object released from rest in the presence of air friction.

Answer: (d)
MC   Moderate

2.26: Suppose that an object is moving with constant acceleration. Which of the following is an accurate statement concerning its motion?
a) In equal times its speed increases by equal amounts.
b) In equal times its velocity changes by equal amounts.
c) In equal times it moves equal distances.
d) None of the above is true.

Answer: (b)
MC   Moderate

2.27: A can, after having been given a kick, moves up along a smooth hill of ice. It will
  a) travel at constant velocity.
  b) have a constant acceleration up the hill, but a different constant acceleration when it comes back down the hill.
  c) have the same acceleration, both up the hill and down the hill.
  d) have a varying acceleration along the hill.

Answer: (c)
  MC Difficult

2.28: When a ball is thrown straight up, its acceleration at the top is zero.

Answer: False
  TF Easy

2.29: Which graph represents an object at rest?

a)

b)

c)

d)

Answer: (a)
  MC Easy

2.30: Which graph represents constant positive acceleration?

a)

b)

c)

d)

Answer: (d)
  MC Moderate

2.31: Under what condition is average velocity equal to the average of the object's initial and final velocity?
   a) The acceleration must be constantly changing.
   b) The acceleration must be constant.
   c) This can only occur if there is no acceleration.

Answer: (b)
   MC   Moderate

2.32: When an object is released from rest and falls in the absence of friction, which of the following is true concerning its motion?
   a) Its acceleration is constant.
   b) Its velocity is constant.
   c) Neither its acceleration nor its velocity is constant.
   d) Both its acceleration and its velocity are constant.

Answer: (a)
   MC   Easy

2.33: A skydiver jumps from a high-flying plane. When she reaches terminal velocity, her acceleration
   a) is essentially zero.
   b) is in the upward direction.
   c) is approximately 9.8 m/s^2 downward.

Answer: (a)
   MC   Moderate

2.34: Which of the following speeds is greatest?
   a) 10 km/h   b) 10 mi/h   c) 10 m/s   d) 10 ft/s

Answer: (c)
   MC   Moderate

2.35: For this problem, assume that the acceleration of gravity is 10 m/s^2 downward, and that all friction effects can be neglected. A ball is thrown upward at a velocity of 20 m/s. What is its velocity after 3 s?
   a) 10 m/s upward.
   b) 10 m/s downward.
   c) zero
   d) none of the above choices is correct.

Answer: (b)
   MC   Easy

2.36: A motorist travels for 3 h at 80 km/h and 2 h at 100 km/h. What is her average speed for the trip?
   a) 85 km/h   b) 88 km/h   c) 90 km/h   d) 92 km/h

Answer: (b)
   MC   Moderate

2.37: A motorist travels 160 km at 80 km/h and 160 km at 100 km/h. What is the average speed of the motorist for this trip?
  a) 84 km/h   b) 89 km/h   c) 90 km/h   d) 91 km/h

Answer: (b)
  MC   Moderate

2.38: A boat can move at 30 km/h in still water. How long will it take to move 12 km upstream in a river flowing 6 km/h?
  a) 20 min   b) 22 min   c) 24 min   d) 30 min

Answer: (d)
  MC   Moderate

2.39: Consider a boat that can travel with speed V in still water. For which of the following trips will the elapsed time be least?
  a) Boat travels a distance 2d in still water.
  b) Boat travels a distance d upstream and then returns to its starting point.
  c) Boat travels a distance d downstream and then returns to its starting point.
  d) The time is the same in all of the above cases.

Answer: (a)
  MC   Moderate

2.40: In a 400-m relay race the anchorman (the person who runs the last 100 m) for the Trojans can run 100 m in 9.8 s. His rival, the anchorman for the Bruins, can cover 100 m in 10.1 s. What is the largest lead the Bruin runner can have when the Trojan runner starts the final leg of the race, in order that the Trojan runner not lose the race?

Answer: 3 m
  ES   Moderate

2.41: Suppose that a Ferrari and a Porsche begin a race with a moving start, and each moves with constant speed. One lap of the track is 2 km. The Ferrari laps the Porsche after the Porsche has completed 9 laps. If the speed of the Ferrari had been 10 km/h less, the Porsche would have traveled 18 laps before being overtaken. What were the speeds of the two cars?

Answer: Porsche speed = 180 km/h;   Ferrari speed = 200 km/h
  ES   Difficult

2.42: A bat, flying due east at 2 m/s, emits a shriek that is reflected back to it from an oncoming insect flying directly toward the bat at 4 m/s. The insect is 20 m from the bat at the instant the shriek is emitted. Sound travels 340 m/s in air. After what elapsed time does the bat hear the reflected echo?

Answer: 116 ms (milliseconds)
  ES   Difficult

2.43: An airplane increases its speed from 100 m/s to 160 m/s, at the average rate of 15 m/s^2. How much time does it take for the complete increase in speed?
    a) 17.3 s   b) 0.0577 s   c) 4.0 s   d) 0.25 s

Answer: (c)
MC   Easy

2.44: A car traveling 30 m/s is able to stop in a distance d. Assuming the same braking force, what distance does this car require to stop when it is traveling twice as fast?
    a) d   b) $\sqrt{2}$ d   c) 2 d   d) 4 d

Answer: (d)
MC   Easy

2.45: A car decelerates uniformly and comes to a stop after 10 s. The car's average velocity during deceleration was 50 km/h. What was the car's deceleration while slowing down?
    a) 10 km/h-s   b) 8 km/h-s   c) 5 km/h-s   d) 4 km/h-s

Answer: (a)
MC   Easy

2.46: A car traveling 60 km/h accelerates at the rate of 2 m/s^2. How much time is required for the car to reach a speed of 90 km/h?

Answer: 4.17 s
ES   Moderate

2.47: A bullet moving horizontally with a speed of 500 m/s strikes a sandbag and penetrates a distance of 10 cm.
        (a) What is the average acceleration of the bullet?
        (b) How long does it take to come to rest?

Answer: (a) 1.25 x 10^6 m/s^2
        (b) 0.4 ms
ES   Moderate

2.48: In which of the following cases will a car move the greatest distance?
    a) A car with speed v_1 travels with constant speed for time t_1 and then accelerates uniformly for t_2 s to speed v_2.
    b) A car with speed v_1 accelerates uniformly to speed v_2 in t_1 seconds and then travels at speed v_2 for t_2 seconds.
    c) It is not possible to answer this question definitively without knowing numerical values for t_1, t_2, v_1, and v_2.

Answer: (b)
MC   Difficult

2.49: A car with good tires on a dry road can decelerate at about 5 m/s^2 when braking. If the car is traveling at 89.5 km/h (55 mi/h), (a) how long does it take the car to stop under these conditions? (b) how far does the car travel during this time?

Answer: (a) 4.9 s   (b) 60.4 m
        ES Moderate

2.50: In a test carried out by a car manufacturer, a test driver is asked to put on his brakes when a warning light is suddenly flashed on in the roadway ahead. When traveling 25 m/s the driver is able to stop in 98 m, and when traveling 10 m/s he is able to stop in 32.5 m. Assuming the driver's reaction time is the same in each case, and that the rate of deceleration when the brakes are applied is independent of speed, determine (a) the driver's reaction time. (b) the driver's deceleration.

Answer: (a) 2.8 s   (b) 11.2 m/s^2
        ES Difficult

2.51: A jet fighter plane is launched from a catapult on an aircraft carrier. It reaches a speed of 42 m/s at the end of the catapult, and this requires 2 s. Assuming the acceleration is constant, what is the length of the catapult?
        a) 16 m   b) 24 m   c) 42 m   d) 84 m

Answer: (c)
        MC Moderate

2.52: A car starting from rest moves with constant acceleration of 2 m/s^2 for 10 s, then travels with constant speed for another 10 s, and then finally slows to a stop with constant acceleration of -2 m/s^2. How far does it travel?
        a) 200 m   b) 300 m   c) 400 m   d) 500 m

Answer: (c)
        MC Moderate

2.53: A car starts from rest and accelerates uniformly at 3 m/s^2. A second car starts from rest 6 s later at the same point and accelerates uniformly at 5 m/s^2. How long does it take the second car to overtake the first car?
        a) 12.2 s   b) 18.9 s   c) 20.6 s   d) 24.0 s

Answer: (c)
        MC Difficult

2.54: A toy rocket is launched upward with a net acceleration of 10 m/s^2 for 3 s. It then slows at the rate of 10 m/s^2 until it reaches its maximum altitude. How high does it go?
        a) 30 m   b) 45 m   c) 60 m   d) 90 m

Answer: (d)
        MC Moderate

2.55: A bullet shot straight up returns to its starting point in 10 s. Its initial speed was
a) 9.8 m/s   b) 24.5 m/s   c) 49 m/s   d) 98 m/s

Answer: (c)
MC   Easy

2.56: A ball is thrown straight up with a speed of 36 m/s. How long does it take to return to its starting point?

Answer: 7.34 s
ES   Easy

2.57: A brick is thrown downward from the top of a building with an initial speed of 25 m/s. It strikes the ground after 2.0 s. How high is the building? (Use $g = 10$ m/s^2.)
a) 20 m   b) 30 m   c) 50 m   d) 70 m

Answer: (d)
MC   Difficult

2.58: A car goes from 40 m/s to 80 m/s in a distance of 200 m. What is its acceleration during this time?
a) 8.0 m/s^2 b) 9.6 m/s^2 c) 12 m/s^2 d) 24 m/s^2

Answer: (c)
MC   Easy

2.59: An object goes from 14 m/s to 4 m/s in 10 s with constant acceleration. The acceleration is
a) -10 m/s^2 b) -1.4 m/s^2 c) -1.0 m/s^2 d) -0.4 m/s^2

Answer: (c)
MC   Easy

2.60: A car drives 8 km at 10 m/s, then 40 km at 25 m/s. What is the average speed?
a) 15 m/s   b) 17.5 m/s   c) 20 m/s   d) 22.5 m/s

Answer: (c)
MC   Difficult

2.61: An object travels at 4 m/s for 25 s and then at 20 m/s for 15 s. The average speed is
a) 8.0 m/s   b) 10.0 m/s   c) 12.0 m/s   d) 15 m/s

Answer: (b)
MC   Difficult

2.62: An object slows from 16 m/s to 4 m/s at a rate of -6 m/s^2.
During this time, how far does the object go?
   a) 8 m   b) 12 m   c) 20 m   d) 32 m

Answer: (c)
   MC   Easy

2.63: A car travels at 15 m/s for 10 s. It then speeds up with a
constant acceleration of 2.0 m.s^2 for 15 s. At the end of this
time, what is its velocity?
   a) 15 m/s   b) 30 m/s   c) 45 m/s   d) 375 m/s

Answer: (c)
   MC   Moderate

2.64: A ball is thrown vertically upward. At its highest point, it is
feeling an acceleration.

Answer: True
   TF   Easy

2.65: A ball is thrown straight up with a speed of 30 m/s. Using g =
10 m/s^2, the maximum height the ball reaches is
   a) 30 m   b) 45 m   c) 90 m   d) 135 m

Answer: (b)
   MC   Easy

2.66: A bicycle is traveling at 17 km/h. The brakes are applied, and
the speed one second later is 9 km/h. The rate of acceleration
is
   a) 8 km/s^2 b) -8 km/s^2 c) 2.2 m/s^2 d) -2.2 m/s^2

Answer: (d)
   MC   Moderate

2.67: If 1 mile = 1.61 km, convert 100 km/h to mi/hr.
   a) 161 mi/h   b) 100 mi/h   c) 62 mi/h   d) 55 mi/h

Answer: (c)
   MC   Easy

2.68: If 1 m^i = 1.61 km, convert 30.0 m/s to mi/h.
   a) 8.33 mi/h   b) 18.6 mi/h   c) 67.1 mi/h   d) 242 mi/h

Answer: (c)
   MC   Moderate

2.69: A car goes 30 km in 15 minutes. What is the car's average speed?
a) 2 km/h   b) 15 km/h   c) 120 km/h   d) 450 km/h

Answer: (c)
MC   Easy

2.70: A plane flies in a straight line for 1800 km. It travels the first half of the distance at 200 km/h and the second half at 300 km/h. What is its average speed?
a) 200 km/h   b) 240 km/h   c) 250 km/h   d) 300 km/h

Answer: (b)
MC   Moderate

2.71: An object moves in a straight line in the +x direction. It goes from 10 m/s to 4 m/s in 2 s. What is its average acceleration during this time?
a) -2 m/s^2 b) -3 m/s^2 c) -5 m/s^2 d) -7 m/s^2

Answer: (b)
MC   Easy

2.72: A polar bear starts at the North pole. It travels 1 km South, then 1 km East, then 1 km North to return to its starting point. This trip takes 30 min. What was the bear's average speed?
a) 0 km/h   b) 0.10 km/h   c) 3 km/h   d) 6 km/h

Answer: (d)
MC   Easy

2.73: A polar bear starts at the North Pole. It travels 1 km South, then 1 km East, then 1 km North to return to its starting point. This trip takes 30 min. What was the bear's average velocity?
a) 0 km/h   b) 0.1 km/h   c) 3 km/h   d) 6 km/h

Answer: (a)
MC   Moderate

2.74: A ball is thrown straight up. What is its instantaneous acceleration just before it reaches its highest point?
a) Zero   b) Slightly less than g.
c) Exactly g.   d) Slightly greater than g.

Answer: (c)
MC   Easy

2.75: An object starts from rest and undergoes uniform acceleration. During the first second it travels 5 m. How far will it move during the next second?
a) 5 m   b) 15 m   c) 20 m   d) 25 m

Answer: (b)
MC   Moderate

2.76: An object starts from rest and undergoes uniform acceleration. During the first second it travels 5 m. How far will it travel during the third second?

a) 5 m   b) 15 m   c) 25 m   d) 45 m

Answer: (c)
MC   Moderate

2.77: A brick is dropped from the top of a building. A second brick is thrown straight down from the same building. They are released at the same time. Neglect air resistance. Compare the accelerations of the two bricks.

a) The first brick accelerates faster.
b) The second brick accelerates faster.
c) The two bricks accelerate at the same rate.
d) It is impossible to determine from the information given.

Answer: (c)
MC   Moderate

2.78: Objects A and B both start at rest. They both accelerate at the same rate. However, object A accelerates for twice the time as object B. Compare the final speed of object A to that of object B.

a) the same speed   b) twice as fast
c) three times as fast   d) four times as fast

Answer: (b)
MC   Moderate

2.79: Objects A and B both start from rest. They both accelerate at the same rate. However, object A accelerates for twice the time as object B. During the times that the objects are being accelerated, compare the distance traveled by object A to that of object B.

a) the same distance   b) twice as far
c) three times as far   d) four times as far

Answer: (d)
MC   Moderate

2.80: An object moves in the +x direction with constant acceleration. It starts out going 15 m/s, and 3 s later it is going 9 m/s. What is its average velocity during this time?

a) 12 m/s   b) 5 m/s   c) -6 m/s   d) -2 m/s

Answer: (a)
MC   Easy

2.81: An object is moving in a straight line with constant acceleration. Initially it is travelling at 16 m/s. Three seconds later it is traveling at 10 m/s. How far does it move during this time?

    a) 26 m   b) 30 m   c) 39 m   d) 48 m

Answer: (c)
    MC   Moderate

2.82: The slope of a position versus time graph gives

    a) average velocity   b) instantaneous velocity
    c) average acceleration   d) instantaneous acceleration

Answer: (b)
    MC   Easy

2.83: The slope of a velocity versus time graph gives

    a) average velocity   b) instantaneous velocity
    c) average acceleration   d) instantaneous acceleration

Answer: (d)
    MC   Easy

2.84: Two objects are thrown from the top of a tall building. One is thrown up, and the other is thrown down, both with the same initial speed. Ignore air resistance, and compare their speeds when they hit the street.

    a) The one thrown up is going faster.
    b) The one thrown down is going faster.
    c) They are going the same speed.
    d) It is impossible to tell from the information given.

Answer: (c)
    MC   Moderate

2.85: Ball A is dropped from the top of a building. One second later, ball B is dropped from the same building. As time progresses, the difference in their speeds

    a) increases.
    b) remains constant.
    c) decreases.
    d) cannot be determined from the information given.

Answer: (b)
    MC   Difficult

2.86: Ball A is dropped from the top of a building. One second later,
ball B is dropped from the same building. As time progresses,
the distance between them
a) increases.
b) remains constant.
c) decreases.
d) cannot be determined from the information given.

Answer: (a)
MC Difficult

2.87: Two balls are thrown straight up. The first ball is thrown with
twice the initial speed of the second. Ignore air resistance.
How much higher will the first ball rise?
a) Twice as far.
b) Three times as far.
c) Four times as far.
d) Impossible to tell without knowing the exact speeds.

Answer: (c)
MC Moderate

2.88: A ball is thrown straight up with an initial speed of 30 m/s.
What is its speed after 4.2 s? (Take g = 10 m/s^2.)
   a) 12 m/s   b) 30 m/s   c) 42 m/s   d) 72 m/s

Answer: (a)
MC Moderate

2.89: What is the meaning of a horizontal line on a plot of x vs. t?
a) This describes an object moving with constant non-zero speed.
b) This describes an object moving with constant non-zero
acceleration.
c) This describes an object at rest.
d) Such a graph has no physical meaning, since it describes an
object moving with infinite speed.

Answer: (c)
MC Easy

2.90: How is motion in the negative x direction represented on an x vs.
t plot?
a) By a curve to the left of the origin.
b) By a curve below the horizontal axis.
c) By a downward sloping curve.
d) Such a motion cannot be shown on a simple x vs. t graph.

Answer: (c)
MC Moderate

2.91: What is the meaning of a horizontal line on a plot of v vs. t?
   a) The object is at rest.
   b) The object is moving at constant speed.
   c) The object is speeding up at a constant rate.
   d) The object is accelerating at a constant non-zero rate.

Answer: (b)
   MC   Easy

2.92: An object starts from rest, and accelerates at a constant 6 m/s^2. After 3 s, how far has it gone?
   a) 3 m   b) 6 m   c) 18 m   d) 27 m

Answer: (d)
   MC   Easy

2.93: An object starts from rest, and accelerates at a constant 6 m/s^2. After 3 s, how fast is it going?
   a) 3 m/s   b) 6 m/s   c) 18 m/s   d) 27 m/s

Answer: (c)
   MC   Easy

2.94: An object starts from rest, and accelerates at a constant 6 m/s^2. After 3 s, what is its acceleration?
   a) 3 m/s^2 b) 6 m/s^2 c) 18 m/s^2 d) 27 m/s^2

Answer: (b)
   MC   Easy

2.95: The slope of an x vs. t plot is acceleration.

Answer: False
   TF   Easy

2.96: The slope of a v vs. t plot is acceleration.

Answer: True
   TF   Easy

2.97: A ball is thrown straight up, reaches a maximum height, then falls to its initial height. As the ball is going up, the velocity and acceleration both point up.

Answer: False
   TF   Easy

2.98: A ball is thrown straight up, reaches a maximum height, and falls to its initial height. At the highest point, the velocity and acceleration are both zero.

Answer: False
TF   Easy

2.99: A ball is thrown straight up, reaches a maximum height, and falls to its initial height. As it comes down, the velocity and acceleration both point down.

Answer: True
TF   Easy

3. Kinematics in Two or Three Dimensions; Vectors

3.1: A 400-m tall tower casts a 600-m long shadow over level ground. At what angle θ is the sun elevated above the horizon?
  a) 33.7°
  b) 41.8°
  c) 48.2°
  d) Can't be found; not enough information.

Answer: (a)
MC   Moderate

3.2: The earth is divided into 24 time zones, in each of which local time differs by 1 hour. At 45° N latitude, what is the width of a time zone? (The earth's radius is 6370 km. 45° N latitude is about as far north as Portland, Oregon, and is halfway between the equator and the north pole.)

Answer: 1179 km
ES   Moderate

3.3: It is possible to have constant speed, but still be accelerating.

Answer: True
TF   Easy

3.1: Match the description to the physical quantity.

3.4: average velocity

Answer: $(\mathbf{v}_O + \mathbf{v})/2$
MA   Easy

3.5: instantaneous velocity

Answer: $\mathbf{v}_O + \mathbf{a}\, t$
MA   Easy

3.6: relative velocity

Answer: $\mathbf{v}_{AB} + \mathbf{v}_{BC}$
MA   Easy

3.7: average acceleration

Answer: $(\mathbf{v} - \mathbf{v}_O)/t$
MA   Easy

3.8: range

Answer: R
     MA   Easy

3.9: trajectory

Answer: parabola
     MA   Easy

3.10: resultant

Answer: **R + S + T**
     MA   Easy

3.11: Which of the following is an accurate statement?
    a) A vector cannot have zero magnitude if one of its components is not zero.
    b) The magnitude of a vector can be less than the magnitude of one of its components.
    c) If the magnitude of vector A is less than the magnitude of vector B, then the x-component of A is less than the x-component of B.
    d) The magnitude of a vector can be positive or negative.

Answer: (a)
     MC   Easy

3.12: Which of the following operations will not change a vector?
    a) Translate it parallel to itself.
    b) Rotate it.
    c) Multiply it by a constant factor.
    d) Add a constant vector to it.

Answer: (a)
     MC   Easy

3.13: If you drive west at 20 km/h, then drive east at 15 km/h, your average velocity will be
    a) 5 km/h east.
    b) 35 km/h west.
    c) 35 km/h east.
    d) 5 km/h west.
    e) cannot be determined; insufficient information.

Answer: (e)
     MC   Moderate

3.14: If the acceleration vector of an object is directed anti-parallel
to the velocity vector,
a) the object is turning.
b) the object is speeding up.
c) the object is slowing down.
d) the object is moving in the negative x-direction.

Answer: (c)
MC   Easy

3.15: Two displacement vectors have magnitudes of 5 m and 7 m,
respectively. When these two vectors are added, the magnitude of
the sum
a) is 12 m.
b) could be as small as 2 m, or as large as 12 m.
c) is 2 m.
d) is larger than 12 m.

Answer: (b)
MC   Easy

3.16: If the acceleration of an object is always directed perpendicular
to its velocity,
a) the object is speeding up.
b) the object is slowing down.
c) the object is turning.
d) this situation would not be physically possible.

Answer: (c)
MC   Moderate

3.17:

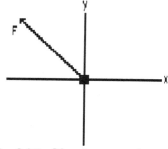

A 200-lb force is pulling on an object, as shown. The sign of
the **x** and **y** components of the force are
a) **x** (positive), **y** (positive).
b) **x** (positive), **y** (negative).
c) **x** (negative), **y** (positive).
d) **x** (negative), **y** (negative).

Answer: (c)
MC   Easy

3.18:

```
        ↑ F1



←
F2




↓ F3
```
                    Three boys each pull with a 20-N force on the same
object.   The resultant force will be
     a) zero.   b) 20 N to the left.
     c) 20 N up.   d) 20 N down.

Answer: (b)
     MC   Easy

38

(3.19) If you want to cross a river in a motorboat in the least amount
       of time, you should head
       a) straight across the river.
       b) slightly upstream.
       c) slightly downstream.
       d) the time will be independent of how you head your boat.

Answer: (a)
     MC   Difficult

3.20: On a calm day (no wind), you can run a 1500-m race at a velocity
      of 4.0 m/s. If you ran the same race on a day when you had a
      constant tailwind which boosts your speed by 2.0 m/s, the time it
      would take you to finish would be
      a) 250 s   b) 750 s   c) 1125 s   d) 9000 s

Answer: (a)
     MC   Easy

3.21: Consider a plane flying with groundspeed V_g and airspeed V_a in a
      wind with velocity V. Which of the following relationships is
      true?
      a) $V_g = V_a + V$
      b) V_g can be greater than $V_a + V$
      c) V_g can be less than $V_a + V$
      d) V_g can have any value between $V_a + V$ and $V_a - V$

Answer: (d)
     MC   Easy

3.22: An airplane with an airspeed of 120 km/h is headed 30° east of
      north in a wind blowing due west at 30 km/h. What is the
      groundspeed of the plane?

Answer: 108 km/h
     ES   Moderate

                              33

3.23: A fighter plane moving 200 m/s in level flight fires a projectile with speed 50 m/s in a forward direction. The projectile makes an angle of 30° with the direction of the plane's motion. What is the speed of the projectile with respect to a stationary observer on the ground?
   a) 245 m/s   b) 250 m/s   c) 268 m/s   d) 293 m/s

Answer: (a)
   MC   Moderate

3.24: A swimmer heading directly across a river 200 m wide reaches the opposite bank in 6 min 40 s. She is swept downstream 300 m. How fast can she swim in still water?
   a) 0.50 m/s   b) 1.24 m/s   c) 1.42 m/s   d) 1.83 m/s

Answer: (a)
   MC   Moderate

3.25: The driver of a motorboat that can move at 10 m/s wishes to travel directly across a narrow strait in which the current flows at 5 m/s. (a) At what angle upstream should the driver head the boat? (b) How long will it take to cross a distance of 1.6 km?

Answer: (a) 26.6°   (b) 179 s
   ES   Moderate

3.26: A butterfly moves with a speed of v = 12.0 m/s. The x-component of its velocity is 8.00 m/s. The angle between the direction of its motion and the x-axis must be
   a) 30.0°   b) 41.8°   c) 48.2°   d) 53.0°

Answer: (c)
   MC   Easy

3.27: An electron, initially moving at 2.0×10^4 m/s along the +x-axis, experiences a constant acceleration of 10^{10} m/s^2 in the +y-direction. How far from its starting point is the electron after 4.0×10^{-6} seconds?
   a) 8.0 m   b) 11.3 m   c) 8.0 cm   d) 11.3 cm

Answer: (d)
   MC   Difficult

3.28: Three forces, each having a magnitude of 30 N, pull on an object in directions that are 120° apart from each other. Which one of the following statements must be true?
   a) The resultant force is zero.
   b) The resultant force is greater than 30 N.
   c) The resultant force is equal to 30 N.
   d) The resultant force is less than 30 N.

Answer: (a)
   MC   Easy

3.29: Two vectors, **R** and **S**, are known:

 and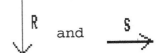

If vector **S** is subtracted from vector **R**, then the vector **T** = **R** - **S** is

a)   b)   c)   d)

Answer: (b)
MC Easy

3.30: Your motorboat can move at 30 km/h in still water. How much time will it take you to move 12 km upstream, in a river flowing at 6 km/h?
a) 20 min   b) 22 min   c) 24 min   d) 30 min

Answer: (d)
MC Easy

3.31: Two vectors **A** and **B** have components (0, 1, 2) and (-1, 3, 1), respectively. (a) Determine the components of the sum of these two vectors. (b) Determine the magnitude of the sum of these two vectors.

Answer: (a) (-1, 4, 3)   (b) 5.1
ES Easy

3.32: Three vectors, expressed in Cartesian coordinates, are

	x-comp	y-comp
S	-3.50	+4.50
T	0	-6.50
U	+5.50	-2.50

The magnitude of the resultant vector S + T + U is
a) 4.92   b) 24.25   c) 16.2   d) 17.5

Answer: (a)
MC Moderate

3.33: If vector **A** = (-3.0 ,- 4.0), and vector **B** = (+3.0 ,-8.0), then the magnitude of vector **C** = **A** - **B** is
a) 13.4   b) 16   c) 144   d) 7.2

Answer: (d)
MC Moderate

3.34: A particle initially moving with a velocity 2 m/s in the +x-direction experiences a constant acceleration of 1 m/s^2 in the x-direction and -2 m/s^2 in the y-direction. What are the velocity components of the particle after 4 s?

Answer: V_x = 6 m/s; V_y = -8 m/s
ES   Easy

3.35: Bullet G is dropped, from rest, at the same time that another bullet H is fired horizontally from a rifle. If both bullets leave from the same height above the ground, then
   a) bullet G hits the ground before bullet H does.
   b) both bullets hit the ground at the same time.
   c) bullet H hits the ground before bullet G does.
   d) not enough information is given to compare when they hit.

Answer: (b)
MC   Easy

3.36: A girl throws a rock horizontally, with a velocity of 10 m/s, from a bridge. It falls 20 m to the water below. How far does the rock travel horizontally before striking the water?(Use g = 10 m/s^2)
   a) 14 m   b) 16 m   c) 20 m   d) 24 m

Answer: (c)
MC   Moderate

3.37: Suppose that several projectiles are fired. Which one will be in the air longest?
   a) The one with the farthest range, R
   b) The one with the highest maximum elevation, h
   c) The one with the greatest initial velocity.

Answer: (b)
MC   Moderate

3.38: An arrow is shot, from a bow, at an original speed of v_o. When it returns to the same horizontal level, its speed will be
   a) ½ v_o b) v_o c) 2 v_o d) 9.8 v_o

Answer: (b)
MC   Moderate

3.39: A golf ball is hit with an initial velocity of 60 m/s at an angle of 30° above the horizontal. How far does it travel?
   a) 152 m   b) 160 m   c) 184 m   d) 318 m

Answer: (d)
MC   Easy

3.40: A package of supplies is dropped from a plane, and one second later a second package is dropped. Neglecting air resistance, the distance between the falling packages will
   a) be constant.   b) decrease.
   c) increase.   d) depend on their weight.

Answer: (c)
   MC Moderate

3.41: A plane flying horizontally at a speed of 50 m/s and at an elevation of 160 m drops a package of supplies. Two seconds later it drops a second package. How far apart will the two packages land on the ground?
   a) 100 m   b) 162 m   c) 177 m   d) 283 m

Answer: (a)
   MC Moderate

3.42: At what angle should a water-gun be aimed in order for the water to land with the greatest horizontal range?
   a) 0°   b) 30°   c) 45°   d) 90°

Answer: (c)
   MC Easy

3.43: The acceleration of gravity on the moon is only one-sixth of that on earth. If you hit a baseball on the moon with the same effort (and at the speed and angle) that you would on earth, the ball would land
   a) the same distance away.   b) one-sixth as far.
   c) 6 times as far.   d) 36 times as far.

Answer: (c)
   MC Easy

3.44: A girl throws a stone straight down from a bridge. The stone leaves her hand with velocity of 8.00 m/s at a height of 12 m above the water below. How much time does it take for the stone to hit the water?
   a) 0.41 s   b) 0.95 s   c) 2.58 s   d) 17.3 s

Answer: (b)
   MC Difficult

3.45: You toss a ball to your friend. When the ball reaches its maximum altitude, its velocity is zero.

Answer: False
   TF Easy

37

3.46: You toss a ball to your friend. When the ball reaches its maximum altitude, its acceleration is zero.

Answer: False
TF   Easy

3.47: You hit a tennis ball over the net. When the ball reaches its maximum height, its speed is
   a) zero.
   b) less than its initial speed.
   c) equal to its initial speed.
   d) greater than its initial speed.

Answer: (b)
MC   Moderate

3.48: You are serving the volleyball for the second time in a volleyball game. If the ball leaves your hand with twice the velocity it had on your first serve, its horizontal range R (compared to your first serve) would be
   a) $\sqrt{2}$ times as much.   b) half as much.
   c) twice as much.   d) four times as much.

Answer: (d)
MC   Easy

3.49: A pilot drops a bomb from a plane flying horizontally. When the bomb hits the ground, the horizontal location of the plane will
   a) be behind the bomb.
   b) be over the bomb.
   c) be in front of the bomb.
   d) depend on the speed of the plane when the bomb was released.

Answer: (b)
MC   Moderate

3.50: A basketball is thrown with a velocity of 20 m/s at an angle of 60° above the horizontal. What is the horizontal component of its instantaneous velocity at the exact top of its trajectory? (Neglect air friction.)
   a) 10 m/s   b) 17.3 m/s   c) 20 m/s   d) zero

Answer: (a)
MC   Moderate

3.51: A soccerball is kicked with a velocity of 25 m/s at an angle of 45° above the horizontal. What is the vertical component of its acceleration as it travels along its trajectory? (Neglect air friction.)
  a) zero
  c) g sin (45°) upward
  e) g cos (45°) downward

  b) g downward
  d) g upward

Answer: (b)
  MC   Easy

3.52: A stone is thrown horizontally from the top of a tower at the same instant that a ball is dropped vertically. Which object is traveling faster when it hits the level ground below?
  a) It is impossible to tell from the information given.
  b) The stone
  c) The ball
  d) Neither, since both are traveling at the same speed.

Answer: (b)
  MC   Moderate

3.53: You hit a handball too hard, and it lands horizontally on top of a roof. If its original velocity was v_0 at an angle θ above the horizontal, then the velocity it lands on the roof with is
  a) zero   b) $v_0 \sin \theta$ c) $v_0 \cos \theta$ d) $2v_0$

Answer: (c)
  MC   Easy

3.54: A ball thrown horizontally from a point 24 m above the ground, strikes the ground after traveling horizontally a distance of 18 m. With what speed was it thrown? (In this problem use g = 10 m/s^2.)
  a) 6.10 m/s   b) 7.40 m/s   c) 8.22 m/s   d) 8.96 m/s

Answer: (c)
  MC   Difficult

3.55: A rifle bullet is fired at an angle of 30° below the horizontal with an initial velocity of 800 m/s from the top of a cliff 80 m high. How far from the base of the cliff does it strike the level ground below?

Answer: 138 m
  ES   Difficult

3.56: A child drops a toy at rest from a point 4 m above the ground at the same instant her friend throws a ball upward at 6 m/s from a point 1 m above the ground. At what distance above the ground do the ball and the toy cross paths?

Answer: 2.78 m
  ES   Difficult

39

3.57: A mortar shell is launched with a velocity of 100 m/s at an angle of 30° above horizontal from a point on a cliff 50 m above a level plain below. How far from the base of the cliff does the mortar shell strike the ground?

Answer: 963 m
ES   Difficult

3.58: In attempting to jump up a waterfall, a salmon leaves the water 2 m from the base of the waterfall. With what minimum speed must it leave the water in order just to make it up a waterfall 1.6 m high?

Answer: 6.60 m/s
ES   Difficult

3.59:

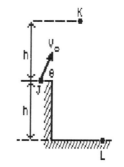

You throw a pebble upward as shown, with a velocity v_O at an angle θ, from a roof (at point J) h ft above the ground. If the pebble rises upward to h ft above the roof (to point K), then falls to the ground below (to point L), its speed when it hits the ground is

a) v_O

b) between v_O and $2v_O$

c) $2v_O$

d) greater than $2v_O$

Answer: (b)
MC   Difficult

3.60: An Olympic athlete throws a javelin at four different angles above the horizontal, each with the same speed: 30° 40° 60° 80°. Which two throws cause the javelin to land the same distance away?

a) 30° and 80°

b) 40° and 60°

c) 40° and 80°

d) 30° and 60°

Answer: (d)
MC   Moderate

3.61: An Air Force plane doing target practice on a floating target sights the target at an angle of 53° from the vertical. The pilot should aim his guns at an angle, from the vertical, of

a) 53°.

b) less than 53°.

c) greater than 53°.

d) the angle depends on the velocity of the plane.

Answer: (c)
MC   Moderate

3.62: Which of the following is an accurate statement?
   a) Three vectors can never add to zero.
   b) If three vectors add to zero, they must be co-linear (i.e., lie in a straight line).
   c) If three vectors add to zero, they must be in a plane.
   d) If three vectors add to zero, they cannot all have the same magnitude.

Answer: (c)
   MC   Easy

3.63: Our second closest star, Alpha Centauri, is 4 ly away. On a particular night, it is located 37° up from the horizon, in a vertical plane. What are its x and y coordinates in this plane (x horizontal, y vertical)?

Answer: x = 3.2 ly,   y = 2.4 ly
   ES   Easy

3.64: A quantity which has a magnitude but no direction is a _____.

Answer: scalar
   ES   Easy

3.65: A quantity which has both magnitude and direction is a _____.

Answer: vector
   ES   Easy

3.66: A bullet is fired horizontally, and at the same instant a second bullet is dropped from the same height. Ignore air resistance. Compare the times of fall of the two bullets.
   a) The fired bullet hits first.
   b) The dropped bullet hits first.
   c) They hit at the same time.
   d) Cannot tell without knowing the masses.

Answer: (c)
   MC   Easy

3.67: Ignoring air resistance, the horizontal component of a projectile's velocity
   a) is zero.   b) remains constant.
   c) continuously increases.   d) continuously decreases.

Answer: (b)
   MC   Easy

3.68: Ignoring air resistance, the vertical component of a projectile's velocity
a) is zero.   b) remains constant.
c) continuously increases.   d) continuously decreases.

Answer: (d)
MC   Easy

3.69: Ignoring air resistance, the horizontal component of a projectile's acceleration
a) is zero.
b) remains a non-zero constant.
c) continuously increases.
d) continuously decreases.

Answer: (a)
MC   Easy

3.70: At its highest point, the acceleration of a projectile is zero.

Answer: False
TF   Easy

3.71: Which of the following statements is accurate?
a) Three vectors can never add to zero.
b) If three vectors add to zero, they must all lie in a straight line.
c) If three vectors add to zero, they must all lie in the same plane.
d) If three vectors add to zero, they cannot all have the same magnitude.

Answer: (c)
MC   Moderate

3.72: Suppose an object has components of acceleration in both the x and y directions.
a) It is moving in a circle.
b) It is moving in a curved path, but not necessarily in a circle.
c) It may be moving in a straight line.
d) Its velocity may be constant.

Answer: (c)
MC   Moderate

3.73: A car travels 20 km West, then 20 km South. What is the magnitude of its displacement?
a) 0 km   b) 20 km   c) 28 km   d) 40 km

Answer: (c)
MC   Easy

3.74: Vector **A** = (1,3). Vector **B** = (3,0). Vector **C** = **A** + **B**. What is the magnitude of **C**?

   a) (4,3)   b) (1,3,3,0)   c) 5   d) 7

Answer: (c)
    MC   Moderate

3.75: Vector **A** has magnitude 8.0 m at an angle of 30 degrees clockwise from the +x axis. The y-conponent of **A** is

   a) 8.0 m   b) 6.9 m   c) 4.0 m   d) -4.0 m

Answer: (d)
    MC   Easy

3.76: A plane is flying on a heading of due South (270°) at 500 km/h. A wind blows from East to West (180°) at 45 km/h. Find the plane's velocity with respect to the ground.

   a) 502 km/h at 265°   b) 502 km/h at 85°
   c) 520 km/h at 5°   d) 545 km/h at 265°

Answer: (a)
    MC   Moderate

3.77: A plane has an air speed of 200 m/s due North, and is in a wind of 50 m/s to the West. The plane's speed relative to the ground is

   a) 150 m/s   b) 200 m/s   c) 206 m/s   d) 250 m/s

Answer: (c)
    MC   Easy

3.1: Match the description to the physical quantity.
   TA

4. Motion and Force: Dynamics

4.1: What is the mass of a woman who weighs 110 lb?
 a) 50 kg   b) 55 kg   c) 110 kg   d) 242 kg

Answer: (a)
    MC   Easy

4.1: Match the unit or expression to the physical quantity.

4.2: SI unit of **a**

Answer: m/s^2
    MA   Easy   Table: 1

4.3: Σ**F** = 0

Answer: equilibrium
    MA   Easy   Table: 1

4.4: SI unit of force

Answer: newton
    MA   Easy   Table: 1

4.5: coefficient of kinetic friction

Answer: dimensionless
    MA   Easy   Table: 1

4.6: the amount of inertia

Answer: mass
    MA   Easy   Table: 1

4.7: You are standing in a moving bus, facing forward, and you
    suddenly fall backward. You can imply from this that the bus's
    a) velocity increased.
    b) velocity decreased.
    c) speed remained the same, but its turning to the right.
    d) speed remained the same, but its turning to the left.

Answer: (a)
    MC   Easy

4.8: Who has a greater weight to mass ratio, a person weighing 400 N or a person weighing 600 N?
a) the person weighing 400 N
b) the person weighing 600 N
c) neither; their ratios are the same
d) the question can't be answered; not enough information is given.

Answer: (c)
MC   Easy

4.9: Its more difficult to start moving a heavy carton from rest than it is to keep pushing it with constant velocity, because
a) the normal force (F_N) is greater when the carton is at rest.
b) $\mu_s < \mu_k$
c) initially, the normal force (F_N) is not perpendicular to the applied force.
d) $\mu_k < \mu_s$

Answer: (d)
MC   Easy

4.10: A packing crate slides down an inclined ramp at constant velocity. Thus we can deduce that
a) a frictional force is acting on it.
b) a net downward force is acting on it.
c) it may be accelerating.
d) it is not acted on by appreciable gravitational force.

Answer: (a)
MC   Easy

4.11: The number of forces acting on a car parked on a hill is
a) one.   b) two.   c) three.   d) four.

Answer: (c)
MC   Moderate

4.12: You fall, while skiing, and one ski (of weight W) loosens and slides down an icy slope (assume no friction), which makes an angle θ with the horizontal. The force that pushes it down along the hill is
a) zero; it moves with constant velocity.
b) W
c) $W \cos \theta$
d) $W \sin \theta$

Answer: (d)
MC   Moderate

4.13: Stacy and David are having a tug-of-war by pulling on opposite ends of a 5-kg rope. Stacy pulls with a 15-N force. What is David's force if the rope accelerates toward Stacy at 2 m/s^2?
a) 3 N   b) 5 N   c) 25 N   d) 50 N

Answer: (b)
MC   Moderate

4.14: When you sit on a chair, the resultant force on you
a) is zero.   b) is up.
c) is down.   d) depends on your weight.

Answer: (a)
MC   Easy

4.15: A book can slide down a frictionless hill at constant velocity.

Answer: False
TF   Moderate

4.16: 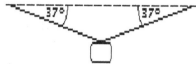 A 1300-N gondola car, at a ski lift, is temporarily suspended at the halfway point, causing the wire to sag by 37° below the horizontal. The tension in the cable is
a) 814 N   b) 1080 N   c) 1628 N   d) 2160 N

Answer: (b)
MC   Moderate

4.17:  A rope, tied between a motor and a crate, can pull the crate up a hill of ice (assume no friction) at constant velocity. The free-body diagram of the crate should contain
a) one force.   b) two forces.
c) three forces.   d) four forces.

Answer: (c)
MC   Moderate

4.18: The coefficients of friction for plastic on wood are μ_s = 0.5 and μ_k = 0.4. How much horizontal force would you need to apply to a 3.0 N plastic calculator to start it moving from rest?
a) 0.15 N   b) 1.2 N   c) 1.5 N   d) 2.7 N

Answer: (c)
MC   Easy

4.19:

If two identical masses are attached by a massless cord
passing over a massless, frictionless pulley of an
Atwood's machine, but at different heights, and then released,
a) the lower mass will go down.
b) the higher mass will go down.
c) the masses will not move.
d) the motion will depend on the amount of the masses.

Answer: (c)
MC   Easy

4.20: The same horizontal force is applied to objects of different
mass. Which of the following graphs illustrates the experimental
results?

a) 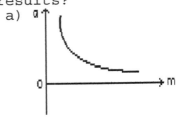   b)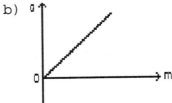

c) 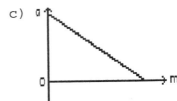   d)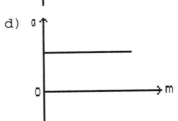

Answer: (a)
MC   Moderate

4.21: If you blow up a balloon, and then release it, the balloon will
fly away. This is an illustration of
a) Newton's First Law.   b) Newton's Second Law.
c) Newton's Third Law.   d) Galileo's Law of Inertia.

Answer: (c)
MC   Easy

4.22: When the rocket engines on the starship NO-PAIN-NO-GAIN are suddenly turned off, while traveling in empty space, the starship will
    a) stop immediately.
    b) slowly slow down, and then stop.
    c) go faster and faster.
    d) move with constant speed.

Answer: (d)
    MC Easy

4.23: If you exert a force F on an object, the force which the object exerts on you will
    a) depend on whether or not the object is moving.
    b) depend on whether or not you are moving.
    c) depend on the relative masses of you and the object.
    d) be F in all cases.

Answer: (d)
    MC Easy

4.24: Batter up! Your bat hits the ball pitched to you with a 1500-N instantaneous force. The ball hits the bat with an instantaneous force, whose magnitude is
    a) somewhat less than 1500 N.
    b) somewhat greater than 1500 N.
    c) exactly equal to 1500 N.
    d) essentially zero.

Answer: (c)
    MC Easy

4.25: A sports car of mass 1000 kg can accelerate from rest to 27 m/s in 7 s. What would the average force of the car's engine be?

Answer: 3860 N
    ES Moderate

4.26: A child's toy is suspended from the ceiling by means of a string. The earth pulls downward on the toy with its weight force of 8 N. If this is the "action force," what is the "reaction force"?
    a) The string pulling upward on the toy with an 8-N force.
    b) The ceiling pulling upward on the string with an 8-N force.
    c) The string pulling downward on the ceiling with an 8-N force.
    d) The toy pulling upward on the earth with an 8-N force.

Answer: (d)
    MC Moderate

4.27: Your mass on the moon will be about one-sixth of your mass on earth.

Answer: False
    TF Easy

4.28: An arrow is shot straight up. At the top of its path, the net
force acting on it is
  a) greater than its weight.
  b) greater than zero, but less than its weight.
  c) instantaneously equal to zero.
  d) equal to its weight.

Answer: (d)
  MC Easy

4.29: A sled of mass 10 kg slides down a flat hill that makes an angle
of 10° with the horizontal. If friction is negligible, what is
the resultant force on the sled?

Answer: 17.0 N
  ES Moderate

4.30: Two toy cars (16 kg and 2 kg) are released simultaneously on an
inclined plane that makes an angle of 30° with the horizontal.
Which statement best describes their acceleration after being
released?
  a) The 16-kg car accelerates 8 times faster than the 2-kg car.
  b) The 2-kg car accelerates 8 times faster than the 16-kg car.
  c) Both cars accelerate at a rate of 0.866 g.
  d) None of the above.

Answer: (d)
  MC Moderate

4.31: A block of mass M slides down a frictionless plane inclined at an
angle θ with the horizontal. The normal reaction force exerted
by the plane on the block is
  a) Mg
  b) Mg sin θ
  c) Mg cos θ
  d) zero, since the plane is frictionless.

Answer: (c)
  MC Easy

4.32: Florence, who weighs 480 N, stands on a bathroom scale in an
elevator. What will she see the scale read when the elevator is
accelerating upward at 4.0 m/s^2?
  a) 480 N   b) 676 N   c) 284 N   d) 196 N

Answer: (b)
  MC Moderate

4.33: A decoration, of mass M, is suspended by a string from the ceiling inside an elevator. The elevator is traveling upward with a constant speed. The tension in the string is
  a) equal to Mg.
  b) less than Mg.
  c) greater than Mg.
  d) impossible to tell without knowing the speed.

Answer: (a)
  MC   Moderate

4.34: A train consists of a caboose (mass = 1000 kg), a car (mass 2000 kg), and an engine car (mass 2000 kg). If the train has an acceleration of 5 m/s^2, then the tension force in the coupling between the middle car and the engine car is
  a) 25,000 N   b) 20,000 N   c) 15,000 N   d) 10,000 N

Answer: (c)
  MC   Moderate

4.35: Two identical masses are attached by a light string that passes over a small pulley, as shown. The table and the pulley are frictionless. The system is moving
  a) with an acceleration less than g.
  b) with an acceleration equal to g.
  c) with an acceleration greater than g.
  d) at constant speed.

Answer: (a)
  MC   Easy

4.36: If you push a 4-kg mass with the same force that you push a 10-kg mass from rest,
  a) the 10-kg mass accelerates 2.5 times faster than the 4-kg mass.
  b) the 4-kg mass accelerates 2.5 times faster than the 10-kg mass.
  c) both masses accelerate at the same rate.
  d) none of the above is true.

Answer: (b)
  MC   Easy

4.37: A 4-kg mass and a 10-kg mass are acted on by the same constant net force during the same amount of time. Both masses are at rest before the force is applied. The 10-kg mass moves a distance X_1 and the 4-kg mass moves a distance X_2 as a result. Which one of the following statements is true?
   a) X_1 is equal to X_2.
   b) The ratio X_1/X_2 is equal to 5/2.
   c) The ratio X_1/X_2 is equal to 2/5.
   d) The ratio X_1/X_2 is equal to $(2/5)^2$.

Answer: (d)
   MC   Moderate

4.38:
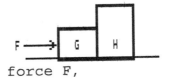
force F,

Two cardboard boxes full of books are in contact with each other on a frictionless table. Box H has twice the mass of box G. If you push on box G with a horizontal force F,

then box H will experience a net force of
   a) 2/3 F   b) F   c) 3/2 F   d) 2 F

Answer: (a)
   MC   Moderate

4.39:
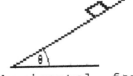
horizontal, for this to occur,

A toolbox, of mass M, is resting on a flat board. One end of the board is lifted up until the toolbox just starts to slide. The angle θ that the board makes with the horizontal, for this to occur, depends on the
   a) mass, M.
   b) acceleration of gravity, g.
   c) normal force.
   d) coefficient of static friction, μ_s.

Answer: (d)
   MC   Difficult

4.40: A horizontal force of 5 N accelerates a 4-kg mass, from rest, at a rate of 0.5 m/s^2 in the positive direction. What friction force acts on the mass?
   a) +3 N   b) -3 N   c) +2 N   d) -2 N

Answer: (b)
   MC   Moderate

4.41: Four students perform an experiment by pulling an object
horizontally across a frictionless table. They repeat the
experiment on several other objects of different masses, always
applying just enough force to produce the same acceleration.
They each graph the results, individually. Which student's graph
is correct?

a) 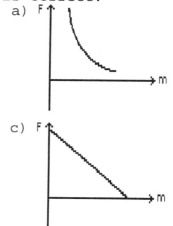   b)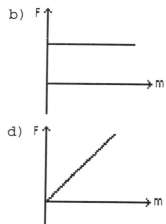

c) F   d) F

Answer: (d)
         MC   Moderate

4.42: A horizontal force accelerates a box from rest across a
horizontal surface (friction is present) at a constant rate. The
experiment is repeated, and all conditions remain the same with
the exception that the horizontal force is doubled. What happens
to the box's acceleration?
     a) It increases to more than double its original value.
     b) It increases to exactly double its original value.
     c) It increases to less than double its original value.
     d) It increases somewhat.

Answer: (a)
         MC   Moderate

4.43: A 2-kg mass and a 5-kg mass are hung on opposite sides of an
Atwood's machine, and released. (a) What is the tension in the
string? (b) How much time is required to fall 60 cms from rest?

Answer: (a) 28 N   (b) 0.53 s
         ES   Difficult

4.44: A load of steel of mass 6000 kg rests on the flatbed of a truck.
It is held in place by metal brackets that can exert a maximum
horizontal force of 8000 N. When the truck is traveling 20 m/s,
what is the minimum stopping distance if the load is not to
slide forward into the cab?

Answer: 150 m
         ES   Difficult

52

4.45: An antitank weapon fires a 3-kg rocket which acquires a speed of 50 m/s after traveling 90 cm down a launching tube. Assuming the rocket was accelerated uniformly, what force acted on it?
  a) 4170 N   b) 3620 N   c) 2820 N   d) 2000 N

Answer: (a)
MC Moderate

4.46: During the investigation of a traffic accident, police find skid marks 90 m long. They determine the coefficient of friction between the car's tires and the roadway to be 0.5 for the prevailing conditions. Estimate the speed of the car when the brakes were applied.

Answer: 29.7 m/s
ES Moderate

4.47: A bulldozer drags a log weighing 500 N along a rough surface. The cable attached to the log makes an angle of 30° with the ground. The coefficient of static friction between the log and the ground is 0.5. What minimum tension is required in the cable in order for the log to begin to move?

Answer: 224 N
ES Difficult

4.48:

      A 16-kg fish is weighed with two spring scales, each of negligible weight, as shown here. What will be the readings on the scales?
  a) Each scale will read 8 kg.
  b) Each scale will read 16 kg.
  c) The top scale will read 16 kg, and the bottom scale will read zero.
  d) The bottom scale will read 16 kg, and the top scale will read zero.
  e) Each scale will show a reading greater than zero and less then 16 kg, but the sum of the two readings will be 16 kg.

Answer: (b)
MC Moderate

4.49: A brick and a feather fall to the earth at their respective
terminal velocities. Which object experiences the greater force
of air friction?
a) The feather
b) The brick
c) Neither, both experience the same amount of air friction.

Answer: (b)
MC   Moderate

 4.50: A wooden block slides directly down an inclined plane, at a
constant velocity of 6 m/s. How much is the coefficient of
kinetic friction (μ_k), if the plane makes an angle of 25° with
the horizontal?
a) 0.47   b) 0.42   c) 0.37   d) 0.91   e) 2.1

Answer: (a)
MC   Moderate

4.51: A flatbed truck carries a load of steel. Only friction keeps the
steel from sliding on the bed of the truck. The driver finds
that the minimum distance in which he can stop from a speed of
8.94 m/s without having the steel slide forward into the truck
cab is 20 m. What would his minimum stopping distance be when
going down a 10° incline?

Answer: 150 m
ES   Difficult

4.52: In the absence of an external force, a moving object will
a) stop immediately.
b) slow down and eventually come to a stop.
c) go faster and faster.
d) move with constant speed.

Answer: (d)
MC   Easy

4.53: Which of the following best expresses the meaning of the word
"force"?
a) potential energy   b) ability to do work
c) push or pull   d) pressure

Answer: (c)
MC   Easy

4.54: A rocket moves through empty space in a straight line with constant speed. It is far from the gravitational effect of any star or planet. Under these conditions, the force that must be applied to the rocket in order to sustain its motion is
a) equal to its weight.
b) equal to its mass.
c) dependent on how fast it is moving.
d) zero.

Answer: (d)
MC   Easy

4.55: A rope one meter long rests on a table where the coefficient of friction is 0.5. What maximum length of rope can hang over the side before the whole rope will slip off?
a) 33 cm
b) 40 cm
c) 50 cm
d) Cannot determine without more information.

Answer: (a)
MC   Moderate

4.56: An example of a force which acts at a distance is
a) tension.   b) weight.
c) static friction.   d) kinetic friction.

Answer: (b)
MC   Moderate

4.57: The force that keeps you from sliding on the sidewalk is
a) weight.   b) kinetic friction.
c) static friction.   d) normal force.

Answer: (c)
MC   Easy

4.58: An object sits on a frictionless surface. A 16 N force is applied to the object, and it accelerates at 2 m/s^2. The mass of the object is
a) 4 kg   b) 8 kg   c) 32 kg   d) 78.4 N

Answer: (b)
MC   Easy

4.59: Starting from rest, a 4 kg body reaches a speed of 8 m/s in 2 s. What is the net force acting on the body?
a) 4 N   b) 8 N   c) 16 N   d) 64 N

Answer: (c)
MC   Moderate

4.2: A 100 kg person has a weight of 1000 N. (Take g = 10 m/s^2.) This person stands on a metric scale inside an elevator. In the next seven questions determine the scale reading for the given situations.

4.60: The elevator is accelerating upward at 0.5 m/s^2.
   a) 950 N
   b) 1000 N
   c) 1050 N
   d) None of the above

Answer: (c)
   MC   Moderate   Table: 2

4.61: The elevator has accelerated upward, and is now moving at constant speed.
   a) 950 N
   b) 1000 N
   c) 1050 N
   d) None of the above

Answer: (b)
   MC   Moderate   Table: 2

4.62: After moving upward at constant speed, the elevator is slowing down at a rate of 0.5 m/s^2.
   a) 950 N
   b) 1000 N
   c) 1050 N
   d) None of the above

Answer: (a)
   MC   Moderate   Table: 2

4.63: The elevator is accelerating downward at 0.5 m/s^2.
   a) 950 N
   b) 1000 N
   c) 1050 N
   d) None of the above

Answer: (a)
   MC   Moderate   Table: 2

4.64: The elevator has accelerated downward and is now moving at constant speed.
   a) 950 N
   b) 1000 N
   c) 1050 N
   d) None of the above

Answer: (b)
   MC   Moderate   Table: 2

4.65: After moving downward at constant speed, the elevator is slowing down at a rate of 0.5 m/s^2.
   a) 950 N
   b) 1000 N
   c) 1050 N
   d) None of the above

Answer: (c)
   MC   Moderate   Table: 2

4.66: The elevator cable breaks, and the automatic brakes fail.
  a) 950 N
  b) 1000 N
  c) 1050 N
  d) None of the above

Answer: (d)
  MC Moderate Table: 2

4.67: Two cars collide head-on. At every moment during the collision, the magnitude of the force the first car exerts on the second is exactly equal to the magnitude of the force the second car exerts on the first. This is an example of
  a) Newton's First Law.
  b) Newton's Second Law.
  c) Newton's Third Law.
  d) no law. This will not happen.

Answer: (c)
  MC Easy

4.68: The same force is applied to the same object on both the Earth and the moon. The acceleration of the object will be the same in both places.

Answer: True
  TF Moderate

4.69: When an object moves with constant velocity, the net force on it is zero.

Answer: True
  TF Easy

4.70: When an object moves in a straight line with decreasing speed, the net force on it must be decreasing.

Answer: False
  TF Moderate

4.71: To hold an 8 N ferret in your hand, you must push upward on it with a force of 8 N. This is an example of Newton's Third Law.

Answer: False
  TF Moderate

4.72: An object of mass m is thrown straight up. At its highest point, the net force on it is
  a) greater than mg.
  b) mg.
  c) less than mg, but greater than zero.
  d) zero.

Answer: (b)
  MC Moderate

4.73: In the absence of all forces, a moving object will
   a) slow down and eventually stop.
   b) move at constant velocity.
   c) immediately come to rest.
   d) Cannot determine without more detail.

Answer: (b)
   MC   Moderate

4.74: Which of the following describes an object in equilibrium?
   a) A car moving with constant velocity down a hill.
   b) The earth orbiting the sun.
   c) A car going around a curve at constant speed.
   d) A ball thrown straight up, at the highest point of its path.

Answer: (a)
   MC   Moderate

4.75: A net force **F** acts on a mass m and produces an acceleration **a**. What acceleration results if a net force 2F acts on mass 4m?
   a) 8**a**   b) 4**a**   c) 2**a**   d) 0.5**a**

Answer: (d)
   MC   Easy

4.76: Mass and weight
   a) both measure the same thing.
   b) are exactly equal.
   c) are two separate quantities.
   d) are both measured in kilograms.

Answer: (c)
   MC   Easy

4.77: The acceleration due to gravity is lower on the moon than on Earth. Which of the following is true about the mass and weight of an astronaut on the moon's surface, compared to Earth?
   a) Mass is less, weight is same.
   b) Weight is less, mass is same.
   c) Both mass and weight are less.
   d) Both mass and weight are the same.

Answer: (b)
   MC   Moderate

4.78: A golf club hits a golf ball with a force of 2400 N. The golf ball hits the club with a force
   a) slightly less than 2400 N.   b) exactly 2400 N.
   c) slightly more than 2400 N.   d) close to 0 N.

Answer: (b)
   MC   Easy

4.79: An object of mass m is hanging by a string from the roof of an elevator. The elevator is moving up at constant speed. What is the tension in the string?
  a) Less than mg.
  b) Exactly mg.
  c) Greater than mg.
  d) Cannot determine without knowing the speed.

Answer: (b)
  MC   Moderate

4.80: An object of mass m is hanging by a string from the roof of an elevator. The elevator is moving upward, but slowing down. What is the tension in the string?
  a) Less than mg.   b) Exactly mg.
  c) Greater than mg.   d) Zero.

Answer: (a)
  MC   Moderate

4.81: A constant net force acts on an object. Describe the motion of the object.
  a) Constant acceleration.   b) Constant speed.
  c) Constant velocity.   d) Increasing acceleration.

Answer: (a)
  MC   Moderate

4.82: An object of mass m sits on a flat table. The Earth pulls on this object with force mg, which we will call the action force. What is the reaction force?
  a) The table pushing up on the object with force mg.
  b) The book pushing down on the table with force mg.
  c) The table pushing down on the floor with force mg.
  d) The object pulling upward on the Earth with force mg.

Answer: (d)
  MC   Moderate

4.83: An object slides on a level surface in the +x direction. It slows and comes to a stop with a constant acceleration of -2.45 m/s^2. What is the coefficient of kinetic friction between the object and the floor?
  a) 0.25
  b) 0.50
  c) -4.9
  d) Impossible to determine without knowing the mass of the object.

Answer: (a)
  MC   Moderate

4.84: Object A weighs 40 N on Earth, and object B weighs 40 N on the moon. The moon's gravity is one sixth of Earth's. Compare the masses of the objects.
   a) A has 6 times the mass of B.
   b) B has 6 times the mass of A.
   c) A and B have equal mass.
   d) The situation as stated is impossible.

Answer: (b)
   MC   Moderate

4.85: A stack of books rests on a level frictionless surface. An unknown force **F** acts on the stack, and it accelerates at 3 m/s^2. A 1 kg book is then added to the stack. The same force is applied, and now the stack accelerates at 2 m/s^2. What was the mass of the original stack?
   a) 1 kg   b) 2 kg
   c) 3 kg   d) None of the above.

Answer: (b)
   MC   Moderate

4.86: A 20 N weight and a 5 N weight are dropped simultaneously from the same height. Ignore air resistance. What happens, and why?
   a) The 20 N weight aceelerates faster because it is heavier.
   b) The 20 N weight accelerates faster because it has more inertia.
   c) They both accelerate at the same rate because they have the same inertia.
   d) The both accelerate at the the same rate because they have the same weight to mass ratio.

Answer: (d)
   MC   Moderate

4.87: A 3 kg object falls toward the ground. Take the weight of the object to be the action force. What is the reaction force?
   a) There is none, because the object is not touching the Earth.
   b) The force of impact when the object hits the ground.
   c) The air resistance puching up on the object.
   d) The object gravitationally pulling upward on the Earth.

Answer: (d)
   MC   Moderate

4.88: Action-reaction forces
   a) sometimes act on the same object.
   b) always act on the same object.
   c) may be at right angles.
   d) always act on different objects.

Answer: (d)
   MC   Moderate

4.89: An object is on a frictionless inclined plane. The plane is inclined at an angle of 30° with the horizontal. What is the object's acceleration?
   a) 0.500 g   b) 0.577 g   c) 0.866 g   d) 1.000 g

Answer: (a)
   MC   Moderate

4.90: An object is placed on an inclined plane. The angle of incline is gradually increased until the onject begins to slide. The angle at which this occurs is θ. What is the coefficient of static friction between the object and the plane?
   a) sin θ
   b) cos θ
   c) tan θ
   d) Cannot determine without knowing the mass of the object.

Answer: (c)
   MC   Moderate

4.1: Match the unit or expression to the physical quantity.
   TA

4.2: A 100 kg person has a weight of 1000 N. (Take g = 10 m/s^2.) This person stands on a metric scale inside an elevator. In the next seven questions determine the scale reading for the given situations.
   TA

5. Circular Motion; Gravitation

5.1: Match the unit to the physical quantity.

5.1: centripetal acceleration

Answer: m/s^2
MA   Easy

5.2: centripetal force

Answer: Newtons
MA   Easy

5.3: universal gravitational constant

Answer: $N \cdot m^2/kg^2$
MA   Easy

5.4: A car, driven around a circle with constant speed, must have zero acceleration.

Answer: False
TF   Easy

5.5: A stone, of mass m, is attached to a strong string and whirled in a vertical circle of radius r. At the exact <u>top</u> of the path the tension in the string is 3 times the stone's weight. The stone's speed at this point is given by
a) $2\sqrt{gr}$ b) $\sqrt{gr}$ c) 4gr   d) $\sqrt{2gr}$

Answer: (a)
MC   Moderate

5.6: The maximum speed around a level curve is 30 km/h. What is the maximum speed around a curve with twice the radius? (Assume all other factors remain unchanged.)
a) 42.4 km/h   b) 45 km/h
c) 60 km/h   d) 120 km/h
e) 21.1 km/h

Answer: (a)
MC   Moderate

5.7: What is the centripetal acceleration of a point on the perimeter of a bicycle wheel of diameter 70 cm when the bike is moving 8 m/s?
   a) 91 m/s^2 b) 183 m/s^2 c) 206 m/s^2 d) 266 m/s^2

Answer: (b)
MC   Moderate

5.8: Two horizontal curves on a bobsled run are banked at the same angle, but one has twice the radius of the other. The safe speed (no friction needed to stay on the run) for the smaller radius curve is V. Therefore, the safe speed on the larger radius curve is
   a) approximately 0.707 V.   b) 2 V.
   c) approximately 1.41 V.   d) 0.5 V.

Answer: (c)
MC   Moderate

5.9: The banking angle in a turn on the Olympic bobsled track is not constant, but increases upward from the horizontal. Coming around a turn, the bobsled team will intentionally "climb the wall," then go lower coming out of the turn. Why do they do this?
   a) To give the team better control, because they are able to see ahead of the turn.
   b) To prevent the bobsled from turning over.
   c) To take the turn at a faster speed.
   d) To reduce the g-force on them.

Answer: (c)
MC   Easy

5.10: A horizontal curve on a bobsled run is banked at a 45° angle. When a bobsled rounds this curve at the curve's safe speed (no friction needed to stay on the run), what is its centripetal acceleration?
   a) 1.0 g   b) 2.0 g
   c) 0.5 g   d) None of the above

Answer: (a)
MC   Moderate

5.11: You can ride your bicycle in a circle while your wheels are perpendicular to the ground.

Answer: False
TF   Easy

5.12: What is the centripetal acceleration of a person at the equator, expressed in a multiple of g?

Answer: 0.0034 g
ES   Moderate

5.13: A car traveling 20 m/s rounds an 80-m radius horizontal curve with the tires on the verge of slipping. How fast can this car round a second curve of radius 320 m? (Assume the same coefficient of friction between the car's tires and each road surface.)
  a) 40 m/s
  b) 30 m/s
  c) 80 m/s
  d) None of the above

Answer: (a)
MC   Easy

5.14: A roller coaster car is on a track that forms a circular loop in the vertical plane. If the car is to just maintain contact with track at the top of the loop, what is the minimum value for its centripetal acceleration at this point?
  a) g downward
  b) 0.5 g downward
  c) g upward
  d) 2 g upward

Answer: (a)
MC   Moderate

5.15: A roller coaster car (mass = M) is on a track that forms a circular loop (radius = r) in the vertical plane. If the car is to just maintain contact with the track at the top of the loop, what is the minimum value for its speed at that point?
  a) 2 Mrg
  b) $\sqrt{2rg}$
  c) $\sqrt{rg}$
  d) None of the above

Answer: (c)
MC   Moderate

5.16: Who was the first person to realize that the planets move in elliptical paths around the sun?
  a) Kepler   b) Brahe   c) Einstein   d) Copernicus

Answer: (a)
MC   Easy

5.17: What is the gravitational force on a 70-kg person, due to the moon? The mass of the moon is 7.36×10^{22} kg and the distance to the moon is 3.82×10^{8} m.

Answer: 0.0024 N
ES   Easy

5.18: The net force on an object in orbit is zero.

Answer: False
TF   Easy

5.19:

A planet revolves clockwise around a star, with constant speed. The direction of its acceleration at point P is

a)   b)   c)   d)

Answer: (d)
MC   Easy

5.20:

A girl attaches a rock to a string, which she then swings counter-clockwise in a horizontal circle. The string breaks at point P on the sketch, which shows a bird's-eye view (i.e., as seen from above). What path will the rock follow?

a) A   b) B   c) C   d) D   e) E

Answer: (b)
MC   Easy

5.21: For a spacecraft going from the earth toward the sun, at what point will the gravitational forces due to the sun and the earth cancel?
earth's mass: $m_s = 5.98 \times 10^{24}$ kg
sun's mass: $m_s = 1.99 \times 10^{30}$ kg
earth-sun distance: $r = 1.50 \times 10^{11}$ m

Answer: 2.60×10^8 m from earth's center
ES   Moderate

5.22: A satellite encircles Mars at a distance above its surface equal to 3 times the radius of Mars. The acceleration of gravity of the satellite, as compared to the acceleration of gravity on the surface of Mars, is
a) zero.   b) the same.
c) one-third as much.   d) one-ninth as much.
e) one-sixteenth as much.

Answer: (e)
MC   Moderate

5.23: The maximum force a pilot can stand is about seven times his weight. What is the minimum radius of curvature that a jet plane's pilot, pulling out of a vertical dive, can tolerate at a speed of 250 m/s?

Answer: 1060 m
ES   Moderate

5.24: By how many newtons does the weight of a 100-kg person change when he goes from sea level to an altitude of 5000 m?

Answer: 1.54 N
ES   Difficult

5.25: A satellite is in a low circular orbit about the earth (i.e, it just skims the surface of the earth). How long does it take to make one revolution around the earth?

Answer: 86.8 min
ES   Moderate

5.26: A pilot executes a vertical dive, then follows a semi-circular arc until it is going straight up. Just as the plane is at its lowest point, the force on him is
a) less than g, and pointing up.
b) less than g, and pointing down.
c) more than g, and pointing up.
d) more than g, and pointing down.

Answer: (c)
MC   Moderate

5.27: A jet plane flying 600 m/s experiences an acceleration of 4 g when pulling out of the dive. What is the radius of curvature of the loop in which the plane is flying? (For this problem use $g = 10 \text{ m/s}^2$.)
a) 640 m   b) 1200 m   c) 7000 m   d) 9000 m

Answer: (d)
MC   Easy

5.28: A pilot makes an outside vertical loop (in which the center of the loop is beneath him) of radius 3200 m. At the top of his loop he is pushing down on his seat with only one-half of his normal weight. How fast is he going?

Answer: 125 m/s
ES   Easy

5.29: A spaceship is traveling to the moon. At what point is it beyond
the pull of earth's gravity?
a) When it gets above the atmosphere
b) When it is half-way there
c) When it is closer to the moon than it is to earth
d) It is never beyond the pull of earth's gravity.

Answer: (d)
MC   Moderate

5.30: Satellite A has twice the mass of satellite T, and rotates in the
same orbit.
a) The speed of T is twice the speed of A.
b) The speed of T is half the speed of A.
c) The speed of T is one-fourth the speed of A.
d) The speed of T is equal to the speed of A.

Answer: (d)
MC   Easy

5.31: Consider a small satellite moving in a circular orbit (radius r)
about a spherical planet (mass M). Which expression gives this
satellite's orbital velocity?
a) v = GM/r   b) $\sqrt{GM/r}$

c) $v = \sqrt{GM/r^2}$ d) $\sqrt{Gr}$

Answer: (b)
MC   Easy

5.32: A spherically symmetric planet has four times the earth's mass
and twice its radius. If a jar of peanut butter weighs 12 N on
the surface of the earth, how much would it weigh on the surface
of this planet?
a) 6 N   b) 12 N
c) 24 N   d) None of the above

Answer: (b)
MC   Moderate

5.33: An earth satellite is in circular orbit 230 km above the surface
of the earth. It is observed to have a period of 89 min. From
this information, estimate the mass of the earth.

Answer: 6.0×10^{24} kg
ES   Moderate

5.34: Europa, a moon of Jupiter, has an orbital diameter of 1.34×10^9
m, and a period of 3.55 days. What is the mass of Jupiter?

Answer: 1.89×10^{27} kg
ES   Moderate

5.35: An astronaut goes out for a "space-walk" at a distance above the earth equal to the radius of the earth. Her acceleration will be
    a) zero   b) g   c) 1/2 g   d) 1/4 g

Answer: (d)
    MC   Moderate

5.36: A planet is discovered to orbit around a star in the galaxy Andromeda, with the same orbital diameter as the earth around our sun. If that star has 4 times the mass of our Sun, what will the period of revolution of that new planet be, compared to the earth's orbital period?
    a) One-fourth as much   b) One-half as much
    c) Twice as much   d) Four times as much

Answer: (b)
    MC   Moderate

5.37: The speed of Halley's comet, while traveling in its elliptical orbit around the sun,
    a) is constant.
    b) increases as it nears the sun.
    c) decreases as it nears the sun.
    d) is zero at two points in the orbit.

Answer: (b)
    MC   Moderate

5.38: The following statements refer to man-made, artificial satellites in orbit around earth. Which is an accurate statement?
    a) It is possible to have a satellite traveling at either a high speed or at a low speed in a given circular orbit.
    b) Only circular orbits (and not elliptical ones) are possible for artificial satellites.
    c) A satellite in a large diameter circular orbit will always have a longer period of revolution about the earth than will a satellite in a smaller circular orbit.
    d) The velocity required to keep a satellite in a given orbit depends on the mass of the satellite.
    e) The period of revolution of a satellite moving about the earth is independent of the size of the orbit it travels.

Answer: (c)
    MC   Moderate

5.39: A coin of mass m rests on a turntable a distance r from the axis of rotation. The turntable rotates with frequency f. Derive an expression for the minimum coefficient of friction between the turntable and the coin if the coin is not to slip.

Answer: $(4\pi^2 r\, f^2)/g$
    ES   Moderate

5.40: An astronaut is inside a space capsule in orbit around the earth. She is able to float inside the capsule because
a) her weight is zero, and her capsule's weight is zero.
b) her weight is zero, and her capsule is accelerating.
c) she and her capsule move with the same constant velocity.
d) she and her capsule move with the same constant acceleration.

Answer: (d)
MC   Moderate

5.41: A race car travels once around a circular track at constant speed.
a) It has constant non-zero acceleration.
b) Its average acceleration is zero.
c) Its instantaneous acceleration is independent of its speed.
d) Its instantaneous acceleration is independent of the size of the track.

Answer: (b)
MC   Moderate

5.42: Consider a particle moving with constant speed such that its acceleration is always perpendicular to its velocity.
a) It is moving in a straight line.
b) It is moving in a circle.
c) It is moving in a parabola.
d) None of the above is definitely true all of the time.

Answer: (d)
MC   Moderate

5.43: Is it possible for an object moving around a circular path to have both centripetal and tangential acceleration?
a) No, because then the path would not be a circle.
b) No, an object can only have one or the other at any given time.
c) Yes, this is possible if the speed is constant.
d) Yes, this is possible if the speed is changing.

Answer: (d)
MC   Moderate

5.44: An object moving in a circular path experiences
a) free fall.   b) constant acceleration.
c) linear acceleration.   d) centripetal acceleration.

Answer: (d)
MC   Easy

5.45: An object moves in a circular path at a constant speed. Consider the direction of the object's velocity and acceleration vectors.
   a) Both vectors point in the same direction.
   b) The vectors point in opposite directions.
   c) The vectors are perpendicular.
   d) The question is meaningless, since the acceleration is zero.

Answer: (c)
MC   Moderate

5.46: A car goes around a 150 m radius curve with a constant speed of 30 m/s. Its acceleration is
   a) 0 m/s^2 b) 0.17 m/s^2 c) 5 m/s^2 d) 6 m/s^2

Answer: (d)
MC   Easy

5.47: What force is needed to make an object move in a circle?
   a) kinetic friction   b) static friction
   c) centripetal force   d) weight

Answer: (c)
MC   Easy

5.48: A car goes around a curve of radius R at a constant speed v. Then it goes around a curve of radius 2R at speed v. What is the centripetal force on the car as it goes around the second curve, compared to the first?
   a) one-fourth as big   b) one-half as big
   c) twice as big   d) four times as big

Answer: (b)
MC   Easy

5.49: A car goes around a curve of radius R at a constant speed v. Then it goes around a curve of radius 2R at speed 2v. What is the centripetal acceleration on the car as it goes around the second curve, compared to the first?
   a) one fourth as big   b) one half as big
   c) twice as big   d) four times as big

Answer: (c)
MC   Easy

5.50: Suppose a satellite were orbiting the Earth just above the surface. (Ignore air resistance and mountains.) Its centripetal acceleration would be
   a) smaller than g.
   b) equal to g.
   c) larger than g.
   d) Impossible to say without knowing the mass.

Answer: (b)
MC   Moderate

5.51: A car goes around a curve of increasing radius. The centripetal force on the car decreases.

Answer: True
TF   Easy

5.52: If a car goes around a curve at half the speed, the centripetal force on the car is cut in half.

Answer: False
TF   Easy

5.53: As a rocket moves away from the Earth's surface, the rocket's weight decreases.

Answer: True
TF   Moderate

5.54: The mass of an object on the moon is less than its mass on the Earth.

Answer: False
TF   Easy

5.55: A car goes around a curve of radius R at a constant speed v. What is the direction of the net force on the car?
   a) toward the curve's center
   b) away from the curve's center
   c) toward the front of the car
   d) toward the back of the car

Answer: (a)
MC   Moderate

5.56: A car is moving with a constant speed v around a level curve. The coefficient of friction between the tires and the road is 0.4. What is the minimum radius of the curve if the car is to stay on the road?
   a) $0.4v^2/g$
   b) v^2/g
   c) $2.5v^2/g$
   d) Insufficient information is provided to solve.

Answer: (c)
MC   Moderate

5.57: A curve of radius 80 m is banked at 45°. Suppose that an ice storm hits, and the curve is effectively frictionless. What is the safe speed with which to take the curve without either sliding up or down?
a) 9.39 m/s
b) 28 m/s
c) 784 m/s
d) The curve cannot be taken safely.

Answer: (b)
MC Moderate

5.58: A car of mass m goes around a banked curve of radius r with speed v. If the road is frictionless due to ice, the car can still negotiate the curve if the horizontal component of the normal force on the car from the road is equal in magnitude to
a) 0.5 mg
b) mg
c) mv^2/r
d) $\tan[v^2/(rg)]$

Answer: (c)
MC Moderate

5.59: A frictionless curve of radius 100 m, banked at an angle of 45°, may be safely negotiated at a speed of
a) 22.1 m/s   b) 31.3 m/s   c) 44.3 m/s   d) 67.1 m/s

Answer: (b)
MC Moderate

5.60: Two objects attract each other gravitationally. If the distance between their centers doubles, the gravitational force
a) quadruples.
b) doubles.
c) is cut in half.
d) is cut to a fourth.

Answer: (d)
MC Easy

5.61: Two objects gravitationally attract each other. If the distance between their centers is cut in half, the gravitational force between them
a) quadruples.
b) doubles.
c) is cut in half.
d) is cut to one fourth.

Answer: (a)
MC Easy

5.62: Two objects, with masses m_1 and m_2, are originally a distance r apart. The gravitational force between them has magnitude F. The second object has its mass changed to $2m_2$, and the distance is changed to r/4. What is the magnitude of the new gravitational force?
  a) F/32   b) F/16   c) 16F   d) 32F

Answer: (d)
  MC   Moderate

5.63: Two objects, with masses m_1 and m_2, are originally a distance r apart. The magnitude of the gravitational force between them is F. The masses are changed to $2m_1$ and $2m_2$, and the distance is changed to 4r. What is the magnitude of the new gravitational force?
  a) F/32   b) F/16   c) 16F   d) 32F

Answer: (b)
  MC   Moderate

5.64: An object weighs 432 N on the surface of the earth. The earth has radius R. If the object is raised to a height of 3R (above the earth's surface), what is its weight?
  a) 432 N   b) 48 N   c) 27 N   d) 0 N

Answer: (c)
  MC   Moderate

5.65: Two moons orbit a planet in nearly circular orbits. Moon A has orbital radius R, and moon B has orbital radius 4R. Moon A takes 20 days to complete one orbit. How long does it take moon B to complete an orbit?
  a) 20 days   b) 80 days   c) 160 days   d) 320 days

Answer: (c)
  MC   Moderate

5.66: At a distance of 14000 km from some planet's center (and above its surface), the acceleration of gravity is 32 m/s^2. What is the acceleration of gravity at a point 28000 km from the planet's center?
  a) 8 m/s^2
  b) 16 m/s^2
  c) 128 m/s^2
  d) Cannot be determined from the information given.

Answer: (a)
  MC   Moderate

5.67: At a distance of 14000 km above some planet's surface, the acceleration of gravity is 32 m/s^2. What is the acceleration of gravity at a point 28000 km above the planet's surface?
   a) 8 m/s^2
   b) 16 m/s^2
   c) 128 m/s^2
   d) Cannot be determined from the information given.

Answer: (d)
   MC   Moderate

5.68: Who was the first to determine that the planets orbit the sun in elliptical orbits with the sun at one focus?
   a) Brahe   b) Kepler   c) Galileo   d) Newton

Answer: (b)
   MC   Moderate

5.69: Let the average orbital radius of a planet be R. Let the orbital period be T. What quantity is constant for all planets orbiting the sun?
   a) T/R   b) T/R^2 c) T^2/R^3 d) T^3/R^2

Answer: (c)
   MC   Moderate

5.70: Two planets have the same surface gravity, but planet B has twice the mass of planet A. If planet A has radius r, what is the radius of planet B?
   a) 0.707r   b) r   c) 1.41r   d) 4r

Answer: (c)
   MC   Moderate

5.71: Two planets have the same surface gravity, but planet B has twice the radius of planet A. If planet A has mass m, what is the mass of planet B?
   a) 0.707m   b) m   c) 1.41m   d) 4m

Answer: (d)
   MC   Moderate

5.72: A motorcycle has a mass of 250 kg. It goes around a 13.7 m radius turn at 96.54 km/h. What is the centripetal force on the motorcycle?
   a) 719 N   b) 2948 N   c) 13122 N   d) 43146 N

Answer: (c)
   MC   Moderate

5.73: A car goes around a flat curve of radius 50 m at a speed of 14 m/s. What must be the minimum coefficient of friction between the tires and the road for the car to make the turn? (Use g = 10 m/s^2.)

a) 0.196
b) 0.382
c) 0.784
d) Cannot determine without knowing the mass of the car.

Answer: (c)
MC   Moderate

5.74: The acceleration of gravity on the moon is one-sixth what it is on earth. An object of mass 72 kg is taken to the moon. What is its mass there?

a) 12 kg   b) 72 kg   c) 72 N   d) 117.6 kg

Answer: (b)
MC   Moderate

5.75: The radius of the earth is R. At what distance above the earth's surface will the acceleration of gravity be 4.9 m/s^2?

a) 0.41 R   b) 0.50 R   c) 1.00 R   d) 1.41 R

Answer: (a)
MC   Moderate

5.76: The earth has radius R. A satellite of mass 100 kg is at a point 3R above the earth's surface. What is the satellite's weight?

a) 62 N   b) 110 N   c) 9000 N   d) 16000 N

Answer: (a)
MC   Moderate

5.77: The acceleration of gravity on the moon is one-sixth what it is on earth. The radius of the moon is one-fourth that of the earth. What is the moon's mass compared to the earth's?

a) 1/6   b) 1/16   c) 1/24   d) 1/96

Answer: (d)
MC   Moderate

5.78: The average distance from the earth to the sun is defined as one "astronomical unit" (AU). An asteroid orbits the sun in one-third of a year. What is the asteroid's average distance from the sun?

a) 0.192 AU   b) 0.48 AU   c) 2.08 AU   d) 5.20 AU

Answer: (b)
MC   Moderate

5.1: Match the unit to the physical quantity.
TA

6. Work and Energy

6.1: Match the unit or expression to the physical quantity.

6.1: joule

Answer: N·m
MA   Easy

6.2: kinetic energy

Answer: ½ mv^2
MA   Easy

6.3: potential energy

Answer: mgy
MA   Easy

6.4: conservative force

Answer: weight
MA   Easy

6.5: non-conservative force

Answer: friction
MA   Easy

6.6: spring constant

Answer: N/m
MA   Easy

6.7: hp

Answer: 746 W
MA   Easy

6.8: Hooke's Law

Answer: -kx
MA   Easy

6.9: watt

Answer: J/s
MA   Easy

6.10: Two men, Joel and Jerry, push against a wall. Jerry stops after
10 min, while Joel is able to push for 5 min longer. Compare the
work they do.
a) Joel does 50% more work than Jerry.
b) Jerry does 50% more work than Joel.
c) Joel does 75% more work than Jerry.
d) Neither of them do any work.

Answer: (d)
MC   Easy

6.11: You lift a 10 N physics book up in the air a distance of 1 m, at
a constant velocity of 0.5 m/s. The work done by gravity is
a) +10 J   b) -10 J   c) +5 J   d) -5 J   e) zero

Answer: (b)
MC   Easy

6.12: The area under the curve, on a Force-position (F-x) graph,
represents
a) work.   b) kinetic energy.
c) power.   d) efficiency.

Answer: (a)
MC   Easy

6.13: Which of the following graphs illustrates Hooke's Law?

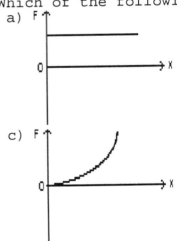
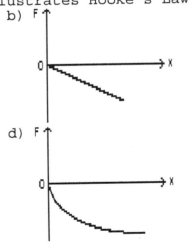

Answer: (b)
MC   Easy

6.14: A 10-kg mass, hung onto a spring, causes the spring to stretch 2
cm. The spring constant is
a) 49 N/cm   b) 5 N/cm   c) 0.2 N/cm   d) 0.02 N/cm

Answer: (a)
MC   Moderate

6.15: Car J moves twice as fast as car K, and car J has half the mass of car K. The kinetic energy of car J, compared to car K is
   a) the same.   b) 2 to 1.   c) 4 to 1.   d) 1 to 2.

Answer: (b)
   MC   Moderate

6.16: An acorn falls from a tree. Compare its kinetic energy K, to its potential energy U.
   a) K increases, and U decreases.
   b) K decreases, and U decreases.
   c) K increases, and U increases.
   d) K decreases, and U increases.

Answer: (a)
   MC   Easy

6.17:

A simple pendulum, consisting of a mass m and a string, swings upward, making an angle θ with the vertical. The work done by the tension force is
   a) zero   b) mg   c) mg cos θ d) mg sin θ

Answer: (a)
   MC   Moderate

6.18: Matthew pulls his little sister Sarah in a sled on an icy surface (assume no friction), with a force of 60 N at an angle of 37° upward from the horizontal. If he pulls her a distance of 12 m, the work he does is
   a) 185 J   b) 433 J   c) 575 J   d) 720 J

Answer: (c)
   MC   Moderate

6.19: A driver, traveling at 22 m/s, slows down her 2000 kg car to stop for a red light. What work is done by the friction force against the wheels?

Answer: 4.84×10^5 J
   ES   Moderate

6.20: A 4.0-kg box of fruit slides 8.0 m down a ramp, inclined at 30° from the horizontal. If the box slides at a constant velocity of 5 m/s, the work done by gravity is
   a) zero   b) +78 J   c) -78 J   d) +157 J   e) -157 J

Answer: (d)
   MC   Moderate

6.21: A 200-g mass attached to the end of a spring causes it to stretch 5.0 cm. If another 200-g mass is added to the spring, the potential energy of the spring will be
   a) the same.
   b) twice as much.
   c) 3 times as much.
   d) 4 times as much.

Answer: (d)
   MC   Moderate

6.22: You throw a ball straight up. Compare the sign of the work done by gravity while the ball goes up with the sign of the work done by gravity while it goes down.
   a) Work up is +, and the work down is +.
   b) Work up is +, and the work down is -.
   c) Work up is -, and the work down is +.
   d) Work up is -, and the work down is -.

Answer: (c)
   MC   Easy

6.2:

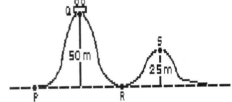

(NOTE: The following question(s) refers to the Cyclone, the famous roller coaster ride at Coney Island, shown in the sketch. (Assume no friction.)

6.23: How much work was required to bring the 1000-kg roller coaster from point P to rest at point Q at the top of the 50 m peak?
   a) 32,000 J   b) 50,000 J   c) 245,000 J   d) 490,000 J

Answer: (d)
   MC   Moderate   Table: 2

6.24: If the roller coaster leaves point Q from rest, how fast is it traveling at point R?
   a) 22.1 m/s   b) 31.3 m/s   c) 490 m/s   d) 980 m/s

Answer: (b)
   MC   Moderate   Table: 2

6.25: If the roller coaster leaves point Q from rest, what is its speed at point S (at the top of the 25-m peak) compared to its speed at point R?
   a) zero   b) $1/\sqrt{2}$ c) $\sqrt{2}$ d) 2   e) 4

Answer: (b)
   MC   Moderate   Table: 2

6.26: A container of water is lifted vertically 3 m, then returned to its original position. If the total weight is 30 N , how much work was done?

 a) 45 J   b) 90 J
 c) 180 J   d) No work was done.

Answer: (d)
    MC   Moderate

6.27: On a plot of F vs. x, what represents the work done by the force F?

 a) The slope of the curve
 b) The length of the curve
 c) The area under the curve
 d) The product of the maximum force times the maximum x

Answer: (c)
    MC   Easy

6.28: A 1-kg flashlight falls to the floor. At the point during its fall, when it is 0.70 m above the floor, its potential energy exactly equals its kinetic energy. How fast is it moving?

 a) 3.7 m/s   b) 6.9 m/s   c) 13.7 m/s   d) 44.8 m/s

Answer: (a)
    MC   Moderate

6.29: A truck weighs twice as much as a car, and is moving at twice the speed of the car. Which statement is true about the truck's kinetic energy (K) compared to that of the car?

 a) All that can be said is that the truck has more K.
 b) The truck has twice the K of the car.
 c) The truck has 4 times the K of the car.
 d) The truck has 8 times the K of the car.

Answer: (d)
    MC   Easy

6.30: Is it possible for a system to have negative potential energy?

 a) Yes, as long as the total energy is positive.
 b) Yes, since the choice of the zero of potential energy is arbitrary.
 c) No, because the kinetic energy of a system must equal its potential energy.
 d) No, because this would have no physical meaning.

Answer: (b)
    MC   Easy

6.31: 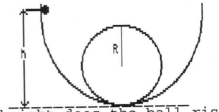 (LOOP-THE-LOOP) A ball is released, from rest, at the left side of the loop-the-loop, at the height shown. If the radius of the loop is R, what vertical height does the ball rise to, on the other side, neglecting friction?

    a) Less than R   b) R
    c) 2R   d) Greater than 2R

Answer: (d)
    MC Moderate

6.32: A spring is characterized by a spring constant of 60 N/m. How much potential energy does it store, when stretched by 1 cm?
    a) 0.003 J   b) 0.3 J   c) 60 J   d) 600 J

Answer: (a)
    MC Easy

6.33: You and your friend want to go to the top of the Eiffel Tower. Your friend takes the elevator straight up. You decide to walk up the spiral stairway, taking longer to do so. Compare the gravitational potential energy (U) of you and your friend, after you both reach the top. Assume you and your friend weigh the same.
    a) Your friend's U is greater than your U, because she got to the top faster.
    b) Your U is greater than your friend's U, because you traveled a greater distance in getting to the top.
    c) Both of you have the same amount of potential energy.
    d) It is impossible to tell, since the times and distances are unknown.

Answer: (c)
    MC Easy

6.34: The total mechanical energy of a system
    a) is equally divided between kinetic energy and potential energy.
    b) is either all kinetic energy or all potential energy, at any one instant.
    c) can never be negative.
    d) is constant, only if conservative forces act.

Answer: (d)
    MC Easy

6.35: A skier, of mass 40 kg, pushes off the top of a hill with an initial speed of 4 m/s. Neglecting friction, how fast will she be moving after dropping 10 m in elevation?
   a) 7 m/s   b) 15 m/s   c) 49 m/s   d) 196 m/s

Answer: (b)
MC   Moderate

6.36:  A simple pendulum, consisting of a mass m, is attached to the end of a 1.5 m length of string. If the mass is held out horizontally, and then released from rest, its speed at the bottom is
   a) 5.4 m/s   b) 9.8 m/s   c) 17 m/s   d) 96 m/s

Answer: (a)
MC   Moderate

6.37: A toy rocket, weighing 10 N, blasts straight up from ground level with a kinetic energy of 40 J. At the exact top of its trajectory, its total mechanical energy is 140 J. To what vertical height does it rise?
   a) 10 m   b) 14 m
   c) 24 m   d) None of the above

Answer: (b)
MC   Moderate

6.38: A boy releases his 2-kg toy, from rest, at the top of a sliding-pond inclined at 20° above the horizontal. What will the toy's speed be after sliding 4 m along the sliding-pond? The coefficient of kinetic friction is 0.2.
   a) 2.21 m/s   b) 3.00 m/s   c) 3.48 m/s   d) 5.18 m/s

Answer: (c)
MC   Easy

6.39: A 4-kg mass moving with speed 2 m/s, and a 2-kg mass moving with a speed of 4 m/s, are gliding over a horizontal frictionless surface. Both objects encounter the same horizontal force, which directly opposes their motion, and are brought to rest by it. Which statement best describes their respective stopping distances?
   a) The 4-kg mass travels twice as far as the 2-kg mass before stopping.
   b) The 2-kg mass travels twice as far as the 4-kg mass before stopping.
   c) Both masses travel the same distance before stopping.
   d) The 2-kg mass travels farther, but not necessarily twice as far.

Answer: (b)
MC   Moderate

6.40: King Kong falls from the top of the Empire State Building, through the air (air friction is present), to the ground below. How does his kinetic energy (K) just before striking the ground compare to his potential energy (U) at the top of the building?
   a) K is equal to U.   b) K is greater than U.
   c) K is less than U.   d) It is impossible to tell.

Answer: (c)
MC   Moderate

6.41: You slam on the brakes of your car in a panic, and skid a certain distance on a straight, level road. If you had been traveling twice as fast, what distance would the car have skidded, under the same conditions?
   a) It would have skidded 4 times farther.
   b) It would have skidded twice as far.
   c) It would have skidded $\sqrt{2}$ times farther.
   d) It is impossible to tell from the information given.

Answer: (a)
MC   Moderate

6.42: How many joules of energy are used by a 1 hp motor that runs for 1 hr?

Answer: 2.69×10^6 J
ES   Easy

6.43: A cyclist does work at the rate of 500 W while riding. How much force does her foot push with when she is traveling at 8 m/s?

Answer: 62.5 N
ES   Easy

6.44: At what rate is a 60-kg boy using energy when he runs up a flight of stairs 10 m high, in 8 s?

Answer: 735 W
ES   Easy

6.45: To accelerate your car at a constant acceleration, the car's engine must
   a) maintain a constant power output.
   b) develop ever-decreasing power.
   c) develop ever-increasing power.
   d) maintain a constant turning speed.

Answer: (c)
MC   Moderate

6.46: Compared to yesterday, you did 3 times the work in one-third the time. To do so, your power output must have been
   a) the same as yesterday's power output.
   b) one-third of yesterday's power output.
   c) 3 times yesterday's power output.
   d) 9 times yesterday's power output.

Answer: (d)
MC   Easy

6.47: A brick and a pebble fall from the roof of an apartment building under construction. At some point the brick is moving at a speed of 3 m/s and the pebble's speed is 5 m/s. If both objects have the same kinetic energy, what is the ratio of the brick's mass to the rock's mass?
   a) 25 to 9   b) 5 to 3
   c) 12.5 to 4.5   d) 3 to 5

Answer: (a)
MC   Moderate

6.48: A 30-N stone is dropped from a height of 10 m, and strikes the ground with a velocity of 7 m/s. What average force of air friction acts on it as it falls?
   a) 22.5 N   b) 75 N   c) 225 N   d) 293 N

Answer: (a)
MC   Difficult

6.49: 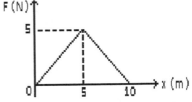 A force that object A exerts on object B is observed over a 10-second interval, as shown on the graph. How much work did object A do during that 10 s?
   a) Zero   b) 12.5 J   c) 25 J   d) 50 J

Answer: (c)
   MC   Moderate

6.50: 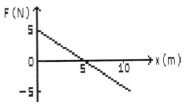 The resultant force you exert on a shopping cart, for a 20 s period, is plotted on the graph, shown. How much work did you do during this 20 s interval?
   a) Zero   b) 12.5 J   c) -25 J   d) 50 J

Answer: (a)
   MC   Moderate

6.51: Does the centripetal force acting on an object do work on the object?
   a) Yes, since a force acts and the object moves, and work is force times distance.
   b) Yes, since it takes energy to turn an object.
   c) No, because the object has constant speed.
   d) No, because the force and the displacement of the object are perpendicular.

Answer: (d)
   MC   Moderate

6.52: The speed needed at the bottom of a loop-the-loop track so that a car can coast to the top, with sufficient speed to stay on the track, depends on the mass of the car. (Neglect the effects of friction.)
   a) Always true.
   b) Never true.
   c) Sometimes true, since kinetic energy is a function of mass.
   d) Sometimes true, since potential energy is a function of mass.

Answer: (b)
   MC   Moderate

6.53: A planet of constant mass orbits the sun in an elliptical orbit. Neglecting any friction effects, what happens to the planet's kinetic energy?
  a) It remains constant.
  b) It increases continually.
  c) It decreases continually.
  d) It increases when the planet approaches the sun, and decreases when it moves farther away.

Answer: (d)
  MC   Moderate

6.54: A spring-driven dart gun propels a 10-g dart. It is cocked by exerting a force of 20 N over a distance of 5 cm. With what speed will the dart leave the gun, assuming the spring has negligible mass?

Answer: 14.1 m/s
  ES   Moderate

6.55: Can work be done on a system if there is no motion?
  a) Yes, if an outside force is provided.
  b) Yes, since motion is only relative.
  c) No, since a system which is not moving has no energy.
  d) No, because of the way work is defined.

Answer: (d)
  MC   Easy

6.56: Does the centripetal force on an object do work on the object?
  a) Yes, since a force acts and the object moves, and work is force times distance.
  b) Yes, since it takes energy to turn an object.
  c) No, because the object has constant speed.
  d) No, because the force and the displacement of the object are perpendicular.

Answer: (d)
  MC   Moderate

6.57: An object is lifted 3 m, then returned to its original position. If the object weighs 40 N, how much work was done on the object?
  a) 240 J   b) 120 J   c) 60 J   d) 0 J

Answer: (d)
  MC   Moderate

6.58: An object is lifted vertically 2 m and held there. If the object weighs 120 N, how much work was done in lifting it?
  a) 480 J   b) 240 J   c) 120 J   d) 0 J

Answer: (b)
  MC   Easy

6.59: On a plot of F vs. x, what represents the work done by the force F?
   a) The slope of the curve.
   b) The length of the curve.
   c) The area under the curve.
   d) The product of the average value of F times the average value of x.

Answer: (c)
   MC   Easy

6.60: A spring has spring constant 60 N/m. If it is stretched by 1 cm, how much potential energy does it store?
   a) 0.003 J   b) 0.006 J   c) 30 J   d) 60 J

Answer: (a)
   MC   Easy

6.61: A 400 N box is pushed up an inclined plane. The plane is 4 m long and rises 2 m. If the plane is frictionless, how much work was done by the push?
   a) 1600 J   b) 800 J   c) 400 J   d) 100 J

Answer: (b)
   MC   Moderate

6.62: A pendulum of length 50 cm is pulled 30 cm away from the vertical axis and released from rest. What will be its speed at the bottom of its swing?
   a) 0.50 m/s   b) 0.79 m/s   c) 1.20 m/s   d) 1.40 m/s

Answer: (d)
   MC   Moderate

6.63: A projectile of mass m leaves the ground with a kinetic energy of 220 J. At the highest point in its trajectory, its kinetic energy is 120 J. To what vertical height, relative to its launch point, did it rise?
   a) 220/mg meters
   b) 120/mg meters
   c) 100/mg meters
   d) Impossible to determine without knowing the angle of launch.

Answer: (c)
   MC   Moderate

6.64: A 2 kg mass is released from rest at the top of a plane inclined at 20° above horizontal. The coefficient of kinetic friction between the mass and the plane is 0.20. What will be the speed of the mass after sliding 4 m along the plane?
   a) 2.21 m/s   b) 3.00 m/s   c) 3.48 m/s   d) 5.18 m/s

Answer: (d)
   MC   Difficult

6.65: An 800 N box is pushed up an inclined plane. The plane is 4 m long and rises 2 m. It requires 3200 J of work to get the box to the top of the plane. What was the magnitude of the average friction force on the box?
   a) 0 N
   b) Non-zero, but less than 400 N
   c) 400 N
   d) Greater than 400 N

Answer: (c)
   MC   Moderate

6.66: A 4 kg mass is moving with speed 2 m/s. A 1 kg mass is moving with speed 4 m/s. Both objects encounter the same constant braking force, and are brought to rest. Which object travels the greater distance before stopping?
   a) the 4 kg mass
   b) the 1 kg mass
   c) both travel the same distance
   d) Cannot be determined from the information given.

Answer: (c)
   MC   Moderate

6.67: Which of the following is not an example of potential energy?
   a) A man on the ledge of a building.
   b) A stretched spring.
   c) A car moving on a flat road.
   d) Water at the top of a waterfall.

Answer: (c)
   MC   Easy

6.68: Describe the energy of a car driving up a hill.
   a) entirely kinetic   b) entirely potential
   c) both kinetic and potential   d) gravitational

Answer: (c)
   MC   Moderate

6.3: For the next three problems, a force F is applied to an object of mass m, moving it at a speed v over a distance d in a time interval t.

6.69: The quantity Fd is
   a) the kinetic energy of the object.
   b) the potential energy of the object.
   c) the work done on the object by the force.
   d) the power supplied to the object by the force.

Answer: (c)
   MC   Easy   Table: 3

6.70: The quantity $1/2\ mv^2$ is
    a) the kinetic energy of the object.
    b) the potential energy of the object.
    c) the work done on the object by the force.
    d) the power supplied to the object by the force.

Answer: (a)
    MC   Easy   Table: 3

6.71: The quantity Fd/t is
    a) the kinetic energy of the object.
    b) the potential energy of the object.
    c) the work done on the object by the force.
    d) the power supplied to the object by the force.

Answer: (d)
    MC   Easy   Table: 3

6.4: For the next three questions, consider a situation in which a 40 N book is carried up a flight of stairs, raising its height by 8 m.

6.72: The work done on the book is
    a) 0 J   b) 5 J   c) 40 J   d) 320 J

Answer: (d)
    MC   Easy   Table: 4

6.73: The change in potential energy of the book is
    a) 0 J   b) 5 J   c) 40 J   d) 320 J

Answer: (d)
    MC   Easy   Table: 4

6.74: The book is dropped. When it reaches its original height, its kinetic energy is
    a) 0 J   b) 5 J   c) 40 J   d) 320 J

Answer: (d)
    MC   Easy   Table: 4

6.75: A person goes up a flight of stairs in 20 s. If the person weighs 600 N, and the vertical height of the stairs is 10 m, the person's power output is
    a) 2 W   b) 300 W   c) 1200 W   d) 120000 W

Answer: (b)
    MC   Moderate

6.76: The energy of a man standing on a ledge is an example of gravitational potential energy.

Answer: True
TF   Easy

6.77: The energy stored in a stretched spring is an example of gravitational potential energy.

Answer: False
TF   Easy

6.78: The energy of a car driving along a level road is an example of kinetic energy.

Answer: True
TF   Easy

6.79: The energy of a car driving up a hill is an example of both kinetic energy and gravitational potential energy.

Answer: True
TF   Easy

6.80: A 50 N object was lifted 2 m vertically and is being held there. How much work is being done in holding the box in this position?
   a) More than 100 J
   b) 100 J
   c) Less than 100 J, but more than 0 J
   d) 0 J

Answer: (d)
MC   Moderate

6.81: Consider two masses m_1 and m_2 at the top of two frictionless inclined planes. Both masses start from rest at the same height. However, the plane on which m_1 sits is at an angle of 30° with the horizontal, while the plane on which m_2 sits is at 60°. If the masses are released, which is going faster at the bottom of its plane?
   a) m_1
   b) m_2
   c) They both are going the same speed.
   d) Cannot be determined without knowing the masses.

Answer: (c)
MC   Moderate

6.82: A ball drops some distance and gains 30 J of kinetic energy. Do not ignore air resistance. How much gravitational potential energy did the ball lose?
a) More than 30 J
b) Exactly 30 J
c) Less than 30 J
d) Cannot be determined from the information given.

Answer: (a)
MC   Moderate

6.83: A ball drops some distance and loses 30 J of gravitational potential energy. Do not ignore air resistance. How much kinetic energy did the ball gain?
a) More than 30 J
b) Exactly 30 J
c) Less than 30 J
d) Cannot be determined from the information given.

Answer: (c)
MC   Moderate

6.84: If you push twice as hard against a stationary brick wall, the amount of work you do
a) doubles
b) is cut in half
c) remains constant but non-zero
d) remains constant at zero.

Answer: (d)
MC   Easy

6.85: A 100 N force has a horizontal component of 80 N and a vertical component of 60 N. The force is applied to a box which rests on a level frictionless floor. The cart starts from rest, and moves 2 m horizontally along the floor. What is the cart's final kinetic energy?
a) 200 J
b) 160 J
c) 120 J
d) Cannot be determined from the information given.

Answer: (b)
MC   Moderate

6.86: An object slides down a frictionless inclined plane. At the bottom, it has a speed of 9.8 m/s. What is the vertical height of the plane?
a) 19.6 m
b) 9.8 m
c) 4.9 m
d) Cannot be determined from the information given.

Answer: (c)
MC   Moderate

6.87: A lightweight object and a very heavy object are sliding with equal speeds along a level frictionless surface. They both slide up the same frictionless hill. Which rises to a greater height?
   a) The heavy object, because it has greater kinetic energy.
   b) The lightweight object, because it weighs less.
   c) They both slide to the same height.
   d) Cannot be determined from the information given.

Answer: (c)
MC   Moderate

6.88: Of the following, which is not a unit of power?
   a) watt/second   b) newton-meter/second
   c) joule/second   d) watt

Answer: (a)
MC   Moderate

6.89: A 12 kg object is moving on a rough, level surface. It has 24 J of kinetic energy. The friction force on it is a constant 0.50 N. How far will it slide?
   a) 2 m   b) 12 m   c) 24 m   d) 48 m

Answer: (d)
MC   Moderate

6.90: An arrow of mass 20 g, is shot horizontally into a bale of hay, striking the hay with a constant velocity of 60 m/s. It penetrates a depth of 20 cm before stopping. What is the average stopping force acting on the arrow?

Answer: 180 N
ES   Moderate

6.1: Match the unit or expression to the physical quantity.
   TA

6.2:

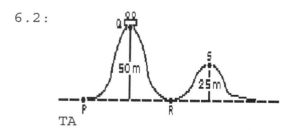

TA

(NOTE: The following question(s) refers to the Cyclone, the famous roller coaster ride at Coney Island, shown in the sketch. (Assume no friction.)

6.1: Match the unit or expression to the physical quantity.

6.3: For the next three problems, a force F is applied to an object of mass m, moving it at a speed v over a distance d in a time interval t.
   TA   Table: 1

6.2:

(NOTE: The following question(s) refers to the Cyclone, the famous roller coaster ride at Coney Island, shown in the sketch. (Assume no friction.)

6.4: For the next three questions, consider a situation in which a 40 N book is carried up a flight of stairs, raising its height by 8 m.
TA   Table: 2

7. Linear Momentum

7.1: Match the physical unit to the physical quantity.

7.1: SI unit of impulse

Answer: N-s
       MA   Difficult

7.2: SI unit of momentum

Answer: kg-m/s
       MA   Easy

7.3: SI unit of $\Delta p / \Delta t$

Answer: newton
       MA   Easy

7.4: metric unit   center of mass

Answer: centimeter
       MA   Easy

7.5: A 1200-kg ferryboat is moving south at 20 m/s. Its momentum is
   a) 1.7×10^{-3} kg·m/s   b) 600 kg·m/s
   c) 2.4×10^3 kg·m/s   d) 2.4×10^4 kg·m/s

Answer: (d)
       MC   Easy

7.6: When a light beach ball rolling with a speed of 6 m/s collides
     with a heavy exercise ball at rest, the beach ball's speed after
     the collision will be, approximately,
   a) 0   b) 3 m/s   c) 6 m/s   d) 12 m/s

Answer: (c)
       MC   Moderate

7.7: In a game of pool, the white cue ball hits the #5 ball and stops,
     while the #5 ball moves away with the same velocity as the cue
     ball had originally. The type of collision is
   a) elastic.
   b) inelastic.
   c) completely inelastic.
   d) any of the above, depending on the mass of the balls.

Answer: (a)
       MC   Moderate

7.8: It is physically impossible to make an instantaneous 90° turn.

Answer: True
   TF   Easy

7.9: Tightrope walkers walk with a long flexible rod in order to
   a) increase their total weight.
   b) allow both hands to hold onto something.
   c) lower their center of mass.
   d) move faster along the rope.

Answer: (c)
   MC   Moderate

7.10: The center of mass of an object must always be located where the physical material of the object is located.

Answer: False
   TF   Easy

7.11: A 2-kg softball is pitched to you at 20 m/s. You hit the ball back along the same path, and at the same speed. If the bat was in contact with the ball for 0.1 s, the average force the bat exerted was
   a) zero   b) 40 N   c) 400 N   d) 800 N

Answer: (d)
   MC   Moderate

7.12: The area under the curve on an F - t graph represents
   a) impulse.   b) momentum.
   c) work.   d) kinetic energy.

Answer: (a)
   MC   Easy

7.13: You (50-kg mass) skate on ice at 4 m/s to greet your friend (40-kg mass), who is standing still, with open arms. As you collide, while holding each other, with what speed do you both move off together?
   a) Zero   b) 2.2 m/s   c) 5 m/s   d) 22.5 m/s

Answer: (b)
   MC   Moderate

7.14: Momentum is conserved during a completely inelastic collision.

Answer: True
   TF   Easy

7.15: A handball of mass 0.1 kg, traveling horizontally at 30 m/s, strikes a wall and rebounds at 24 m/s. What is the change in the momentum of the ball?
  a) 0.6 kg·m/s   b) 1.2 kg·m/s   c) 5.4 kg·m/s   d) 72 kg·m/s

Answer: (c)
  MC   Moderate

7.16: When a cannon fires a cannonball, the cannon will recoil backward because the
  a) energy of the cannonball and cannon is conserved.
  b) momentum of the cannonball and cannon is conserved.
  c) energy of the cannon is greater than the energy of the cannonball.
  d) momentum of the cannon is greater than the energy of the cannonball.

Answer: (b)
  MC   Easy

7.17: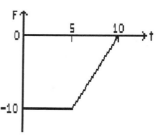
  A force is applied to a 100-kg box, from rest. The force varies with time, as shown on the graph. At t = 10 s, the instantaneous velocity of the box is
  a) Zero
  c) -0.25 m/s
  e) -0.75 m/s
                    b) +0.25 m/s
                    d) +0.75 m/s

Answer: (e)
  MC   Moderate

7.18: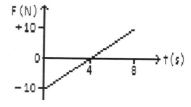
  A force is applied to a 100-kg rock, from rest. The force varies with time, as shown in the graph. At the end of 8 s, the velocity of the rock is
  a) Zero   b) -20 m/s   c) +20 m/s   d) +40 m/s   e) +80 m/s

Answer: (a)
  MC   Moderate

7.19: If you pitch a baseball with twice the kinetic energy you gave it in the previous pitch, the magnitude of its momentum is
a) the same.
b) $\sqrt{2}$ times as much.
c) doubled.
d) 4 times as much.

Answer: (b)
MC   Easy

7.20: Car A (mass = 1000 kg) moves to the right along a level, straight road at a speed of 6 m/s. It collides directly with motorcycle B (mass = 200 kg) in a completely inelastic collision. What is the momentum after the collision if motorcycle B was initially at rest?
a) Zero
b) 6000 kg-m/s to the right
c) 2000 kg-m/s to the right
d) 10,000 kg-m/s to the right.
e) None of the other choices is correct.

Answer: (b)
MC   Moderate

7.21: A proton, of mass m, at rest, is struck head-on by an alpha-particle (which consists of 2 protons and 2 neutrons) moving at velocity +v. If the collision is completely elastic, what speed will the α-particle have after the collision? (Assume the neutron's mass equals the proton's mass.)
a) Zero
b) 2/3 v
c) 3/5 v
d) 5/3 v

Answer: (c)
MC   Difficult

7.22:  A bullet, of mass 20 g, traveling at 350 m/s, strikes a steel plate at an angle of 30° with the plane of the plate. It ricochets off at the same angle, at a speed of 320 m/s. What is the magnitude of the impulse that the wall gives to the bullet?
a) 0.30 N·s
b) 0.52 N·s
c) 6.7 N·s
d) 300 N·s

Answer: (c)
MC   Moderate

7.23: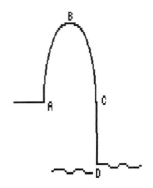

An Olympic diver dives off the high-diving platform. The magnitude of his momentum will be a maximum at point
 a) A.   b) B.   c) C.   d) D.

Answer: (d)
       MC   Easy

7.24: A mine railroad car, of mass 200 kg, rolls with negligible friction on a horizontal track with a speed of 10 m/s. A 70-kg stunt man drops straight down a distance of 4 m, and lands in the car. How fast will the car be moving after this happens?

Answer: 7.41 m/s
       ES   Easy

7.25: A 2000-kg car, traveling to the right at 30 m/s, collides with a brick wall and comes to rest in 0.2 s. The average force the car exerts on the wall is
 a) 12,000 N to the right.   b) 300,000 N to the right.
 c) 60,000 N to the right.   d) none of the above.

Answer: (b)
       MC   Easy

7.26: A 60-kg person walks on a 100-kg log at the rate of 0.8 m/s (with respect to the log). With what speed does the log move, with respect to the shore?

Answer: 0.3 m/s
       ES   Difficult

7.27: A plane, flying horizontally, releases a bomb, which explodes before hitting the ground. Neglecting air resistance, the center of mass of the bomb fragments, just after the explosion
 a) is zero.
 b) moves horizontally.
 c) moves vertically.
 d) moves along a parabolic path.

Answer: (d)
       MC   Moderate

7.28: A ball, of mass 100 g, is dropped from a height of 12 m. Its momentum when it strikes the ground is, in kg·m/s,

    a) 1.5   b) 1.8   c) 2.4   d) 4.8

Answer: (a)
    MC Moderate

7.29: A toy rocket, of mass 120 g, achieves a velocity of 40 m/s after 3 s, when fired straight up. What average thrust force does the rocket engine exert?

    a) 1.2 N   b) 1.6 N   c) 2.8 N   d) 4.4 N

Answer: (c)
    MC Moderate

7.30: A fire hose is turned on the door of a burning building in order to knock the door down. This requires a force of 1000 N. If the hose delivers 40 kg per second, what is the maximum velocity of the stream needed, assuming the water doesn't bounce back?

    a) 15 m/s   b) 20 m/s   c) 40 m/s   d) 50 m/s

Answer: (d)
    MC Moderate

7.31: A sailboat of mass m is moving with a momentum p. Which of the following represents its kinetic energy?

    a) $p^2/2m$ b) $\frac{1}{2} mp^2$ c) mp   d) mp/2

Answer: (a)
    MC Easy

7.32: Some automobiles have air bags installed on their dashboards as a safety measure. The impact of a collision causes the bag to inflate, and it then cushions a passenger's head when he is thrown forward in the car. Suppose that such a bag could lengthen the collision time between one's head and the dashboard by a factor of 10. What effect would this have on the force exerted on the head?

    a) The force would not be reduced, but the energy transferred to the head would be, thereby minimizing the damage done.
    b) The force would be reduced by a factor of about 3.
    c) The force would be reduced by a factor of about 10.
    d) The force would be reduced by a factor of about 100.
    e) The force would be reduced by a factor of about 1000.

Answer: (c)
    MC Easy

7.33: A small bomb, of mass 10 kg, is moving toward the North with a velocity of 4 m/s. It explodes into three fragments: a 5-kg fragment moving west with a speed of 8 m/s; a 4-kg fragment moving east with a speed of 10 m/s; and a third fragment with a mass of 1 kg. What is the velocity of the third fragment? (Neglect air friction.)
   a) Zero
   c) 40 m/s south
   b) 40 m/s north
   d) None of the above

Answer: (b)
   MC   Moderate

7.34: A machine gun, of mass 35 kg, fires 50-gram bullets, with a muzzle velocity of 750 m/s, at the rate of 300 rounds per minute. What is the average force exerted on the machine gun mount?

Answer: 188 N
   ES   Moderate

7.35: A golf ball traveling 3 m/s to the right collides in a head-on collision with a stationary bowling ball in a friction-free environment. If the collision is almost perfectly elastic, the speed of the golf ball immediately after the collision is
   a) slightly less than 3 m/s.
   b) slightly greater than 3 m/s.
   c) equal to 3 m/s.
   d) much less than 3 m/s.

Answer: (a)
   MC   Moderate

7.36: A 100-kg football linebacker moving at 2 m/s tackles head-on an 80-kg halfback running 3 m/s. Neglecting the effects due to digging in of cleats,
   a) the linebacker will drive the halfback backward.
   b) the halfback will drive the linebacker backward.
   c) neither player will drive the other backward.
   d) this is a simple example of an elastic collision.

Answer: (b)
   MC   Moderate

7.37: A small car meshes with a large truck in a head-on collision. Which of the following statements concerning the magnitude of the average collision force is correct?
   a) The truck experiences the greater average force.
   b) The small car experiences the greater average force.
   c) The small car and the truck experience the same average force.
   d) It is impossible to tell since the masses and velocities are not given.

Answer: (c)
   MC   Easy

7.38: Water runs out of a horizontal drainpipe at the rate of 120 kg per minute. It falls 3.2 m to the ground. Assuming the water doesn't splash up, what average force does it exert on the ground?

a) 6.20 N   b) 12.0 N   c) 15.8 N   d) 19.6 N

Answer: (c)
MC   Difficult

7.39: A 2-kg mass moving to the east at a speed of 4 m/s collides head-on in an inelastic collision with a stationary 2-kg mass. How much kinetic energy is lost during this collision?

a) 16 J   b) 4 J   c) 8 J   d) Zero

Answer: (c)
MC   Moderate

7.40: A Ping-Pong ball moving east at a speed of 4 m/s, collides with a stationary bowling ball. The Ping-Pong ball bounces back to the west, and the bowling ball moves very slowly to the east. Which object experiences the greater magnitude impulse during the collision?

a) Neither; both experienced the same magnitude impulse.
b) The Ping-Pong ball
c) The bowling ball
d) It's impossible to tell since the velocities after the collision are unknown.

Answer: (a)
MC   Moderate

7.41: Two equal mass balls (one blue and the other gold) are dropped from the same height, and rebound off the floor. The blue ball rebounds to a higher position. Which ball is subjected to the greater magnitude impulse during its collision with the floor?

a) It's impossible to tell since the time intervals and forces are unknown.
b) Both balls were subjected to the same magnitude impulse.
c) The gold ball
d) The blue ball

Answer: (d)
MC   Moderate

7.42: A 50-gram ball moving +10 m/s collides head-on with a stationary ball of mass 100 g. The collision is elastic. What is the speed of each ball immediately after the collision?

Answer: -3.3 m/s and +6.7 m/s
ES   Difficult

7.43: A car of mass M, traveling with a velocity V, strikes a parked station wagon, whose mass is 2M. The bumpers lock together in this head-on inelastic collision. What fraction of the initial kinetic energy is lost in this collision?
   a) 1/2   b) 1/3   c) 1/4   d) 2/3

Answer: (d)
   MC   Moderate

7.44:

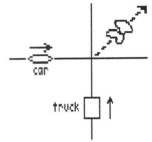

A car (mass = 1500 kg) and a small truck (mass = 2000 kg) collide at right angles at an icy intersection. The car was traveling east at 20 m/s and the truck was traveling north at 20 m/s when the collision took place. What is the speed of the combined wreck, assuming a completely inelastic collision.

Answer: 14.3 m/s
   ES   Moderate

7.45: Three masses are positioned as follows: 2 kg at (0, 0), 2 kg at (2, 0), and 4 kg at (2,1). Determine the coordinates of the center of mass.

Answer: (1.5, 0.5)
   ES   Easy

7.46: Two cars collide head-on on a level friction-free road. The collision was completely inelastic and both cars quickly came to rest during the collision. What is true about the velocity of this system's center of mass?
   a) It was always zero.
   b) It was never zero.
   c) It was not initially zero, but ended up zero.

Answer: (a)
   MC   Moderate

7.47: Consider two unequal masses, M and m. Which of the following statements is _false_?
   a) The center of mass lies on the line joining the centers of each mass.
   b) The center of mass is closer to the larger mass.
   c) It is possible for the center of mass to lie within one of the objects.
   d) If a uniform rod of mass m were to join the two masses, this would not alter the position of the center of mass of the system without the rod present.

Answer: (d)
   MC   Easy

7.48: A 3-kg mass is positioned at (0, 8), and a 1-kg mass is positioned at (12, 0). What are the coordinates of a 4-kg mass which will result in the center of mass of the system of three masses being located at the origin, (0, 0)?

    a) (-3, -6)   b) (-12, -8)   c) (3, 6)   d) (-6, -3)

Answer: (a)
    MC   Moderate

7.49: Two astronauts, of masses 60 kg and 80 kg, are initially at rest in outer space. They push each other apart. What is their separation after the lighter astronaut has moved 12 m?

    a) 15 m   b) 18 m   c) 21 m   d) 24 m

Answer: (c)
    MC   Moderate

7.50: What is the momentum of a 2000 kg truck traveling at 20 m/s?

    a) 100 kg·m/s   b) 19600 kg·m/s
    c) 40000 kg·m/s   d) 400000 kg·m/s

Answer: (c)
    MC   Easy

7.51: Which of the following is an accurate statement?
    a) The momentum of a projectile is constant.
    b) The momentum of a moving object is constant.
    c) If an object is acted on by a non-zero net external force, its momentum will not remain constant.
    d) If the kinetic energy of an object is doubled, its momentum will also double.

Answer: (c)
    MC   Easy

7.52: A ball of mass 0.10 kg is dropped from a height of 12 m. Its momentum when it strikes the ground is
    a) 1.5 kg·m/s   b) 1.8 kg·m/s   c) 2.4 kg·m/s   d) 4.8 kg·m/s

Answer: (a)
    MC   Moderate

7.53: A 2 kg mass moves with a speed of 5 m/s. It collides head-on with a 3 kg mass at rest. If the collision is perfectly inelastic, what is the speed of the masses after the collision?
    a) 10 m/s
    b) 2.5 m/s
    c) 2 m/s
    d) 0 m/s, since the collision is inelastic

Answer: (c)
    MC   Easy

7.54: Which of the following is a false statement?
   a) For a uniform symmetric object, the center of mass is at the center of symmetry.
   b) For an object on the surface of the earth, the center of gravity and the center of mass are the same point.
   c) The center of mass of an object must lie within the object.
   d) The center of gravity of an object may be thought of as the "balance point."

Answer: (c)
MC   Moderate

7.55: A small object collides with a large object and sticks. Which object experiences the larger magnitude of momentum change?
   a) the large object
   b) the small object
   c) Both objects experienece the same magnitude of momentum change.
   d) Cannot be determined from the information given.

Answer: (c)
MC   Moderate

7.56: A small object with a momentum of magnitude 5 kg·m/s approaches head-on a large object at rest. The small object bounces straight back with a momentum of magnitude 4 kg·m/s. What is the magnitude of the small object's momentum change?
   a) 9 kg·m/s   b) 5 kg·m/s   c) 4 kg·m/s   d) 1 kg·m/s

Answer: (a)
MC   Moderate

7.57: A small object with momentum 5 kg·m/s approaches head-on a large object at rest. The small object bounces straight back with a momentum of magnitude 4 kg·m/s. What is the magnitude of the large object's momentum change?
   a) 9 kg·m/s   b) 5 kg·m/s   c) 4 kg·m/s   d) 1 kg·m/s

Answer: (a)
MC   Moderate

7.58: A 3 kg object moves to the right with a speed of 4 m/s. It collides in a perfectly elastic collision with a 6 kg object moving to the left at 2 m/s. What is the total kinetic energy after the collision?
   a) 72 J   b) 36 J   c) 24 J   d) 0 J

Answer: (b)
MC   Moderate

7.59: A 3 kg object moves to the right at 4 m/s. It collides in a perfectly inelastic collison with a 6 kg object moving to the left at 2 m/s. What is the total kinetic energy after the collision?
   a) 72 J   b) 36 J   c) 24 J   d) 0 J

Answer: (d)
   MC   Moderate

7.60: A 3 kg object moves to the right at 4 m/s. It collides head-on with a 6 kg object moving to the left at 2 m/s. Which statement is correct?
   a) The total momentum both before and after the collision is 24 kg·m/s.
   b) The total momentum before the collison is 24 kg·m/s, and after the collision is 0 kg·m/s.
   c) The total momentum both before and after the collision is zero.
   d) None of the above is true.

Answer: (c)
   MC   Moderate

7.61: A constant 6 N net force acts for 4 s on a 12 kg object. What is the object's change of velocity?
   a) 2 m/s   b) 12 m/s   c) 18 m/s   d) 288 m/s

Answer: (a)
   MC   Easy

7.62: A 4 N force acts for 3 s on an object. The force suddenly increases to 15 N and acts for one more second. What impulse was imparted by these forces to the object?
   a) 12 N·s   b) 15 N·s   c) 16 N·s   d) 27 N·s

Answer: (d)
   MC   Moderate

7.63: A very small object moving with speed v collides head-on with a very large object at rest, in a frictionless environment. The collision is almost perfectly elastic. The speed of the large object after the collison is
   a) slightly greater than v.   b) equal to v.
   c) slightly less than v.   d) much less than v.

Answer: (d)
   MC   Moderate

7.64: Two objects move on a level frictionless surface. Object A moves east with a momentum of 24 kg·m/s. Object B moves north with momentum 10 kg·m/s. They make a perfectly inelastic collision. What is the magnitude of their combined momentum after the collision?
   a) 14 kg·m/s
   b) 26 kg·m/s
   c) 34 kg·m/s
   d) Cannot be determined without knowing masses and velocities.

Answer: (b)
   MC   Moderate

7.65: A 4 kg mass sits at the origin, and a 10 kg mass sits at x = + 21 m. Where is the center of mass on the x-axis?
   a) + 7 m   b) + 10.5 m   c) + 14 m   d) + 15 m

Answer: (d)
   MC   Easy

7.66: A rubber ball and a lump of putty have equal mass. They are thrown with equal speed against a wall. The ball bounces back with nearly the same speed with which it hit. The putty sticks to the wall. Which objects experiences the greater momentum change?
   a) The ball.
   b) The putty.
   c) Both experience the same momentum change.
   d) Cannot be determined from the information given.

Answer: (a)
   MC   Moderate

7.67: A very small object moving to the right collides with a very large object at rest. Afterward, the large object moves to the right with a small speed, and the small object moves to the left. Which object experienced the greater magnitude of impulse during the collision?
   a) The large object.
   b) The small object.
   c) Both received the same magnitude of impulse.
   d) Cannot be determined from the information given.

Answer: (c)
   MC   Moderate

7.68: A very large object moving with speed v collides head-on with a very small object at rest. The collision is elastic, and there is no friction. The large object barely slows down. What is the speed of the small object after the collision?
   a) nearly v   b) nearly 2v
   c) nearly 3v   d) nearly infinite

Answer: (b)
   MC   Moderate

7.69: A very large object moving with velocity v collides head-on with a very small object moving with velocity -v. The collision is elastic, and there is no friction. The large object barely slows down. What is the speed of the small object after the collision?
a) nearly v
b) nearly 2 v
c) nearly 3 v
d) nearly infinite

Answer: (c)
MC   Difficult

7.70: A freight car moves along a frictionless level railroad track at constant speed. The car is open on top. A large load of coal is suddenly dumped into the car. What happens to the velocity of the car?
a) It increases.
b) It remains the same.
c) It decreases.
d) Cannot be determined from the information given.

Answer: (c)
MC   Easy

7.71: When is kinetic energy conserved?
a) Elastic collisions.
b) Inelastic collisions.
c) Any collison in which the objects do not stick together.
d) All collisions.

Answer: (a)
MC   Easy

7.72: Two objects collide and bounce off each other. Linear momentum
a) is definitely conserved.
b) is definitely not conserved.
c) is conserved only if the collision is elastic.
d) is conserved only if the environment is frictionless.

Answer: (a)
MC   Moderate

7.73: Two objects collide and bounce off each other. Kinetic energy
a) is definitely conserved.
b) is definitely not conserved.
c) is conserved only if the collision is elastic.
d) is conserved only if the environment is frictionless.

Answer: (c)
MC   Easy

7.74: Two objects collide and stick together. Linear momentum
   a) is definitely conserved.
   b) is definitely not conserved.
   c) is conserved only if the collision is elastic.
   d) is conserved only if the environment is frictionless.

Answer: (a)
   MC   Easy

7.75: Two objects collide and stick together. Kinetic energy
   a) is definitely conserved.
   b) is definitely not conserved.
   c) is conserved only if the collision is elastic.
   d) is conserved only if the environment is frictionless.

Answer: (b)
   MC   Easy

7.76: Kinetic energy is conserved in elastic collisions.

Answer: True
   TF   Easy

7.77: Linear momentum is conserved in all collisions.

Answer: True
   TF   Easy

7.78: Two objects collide and stick together. Linear momentum is
   definitely conserved.

Answer: True
   TF   Easy

7.79: Two objects collide and stick together. Kinetic energy is
   definitely conserved.

Answer: False
   TF   Easy

7.1: Match the physical unit to the physical quantity.
   TA

Rotational Motion

8.1: A wheel of radius 1 m is spinning with a constant angular velocity of 2 rad/s. What is the centripetal acceleration of a point on the wheel's rim?
   a) 0.5 m/s^2 b) 1 m/s^2 c) 2 m/s^2 d) 4 m/s^2

Answer: (d)
   MC   Easy

8.2: A wheel of radius 1 m is spinning with a constant angular velocity of 1 rad/s. What is the translational velocity of a point on the wheel's rim?
   a) 0.5 m/s   b) 1 m/s   c) 2 m/s   d) 4 m/s

Answer: (c)
   MC   Easy

8.3: A boy and a girl are riding on a merry-go-round which is turning at a constant rate. The boy is near the outer edge, and the girl is closer to the center. Who has the greater angular velocity?
   a) The boy
   b) The girl
   c) Both have the same non-zero angular velocity.
   d) Both have zero angular velocity.

Answer: (c)
   MC   Easy

8.4: A boy and a girl are riding on a merry-go-round which is turning at a constant rate. The boy is near the outer edge, and the girl is closer to the center. Who has the greater translational velocity?
   a) The boy.
   b) The girl.
   c) Both have the same non-zero translational velocity.
   d) Both have zero translational velocity.

Answer: (a)
   MC   Easy

8.5: Angular velocity is expressed in units of
   a) meters per second.   b) radians per second.
   c) omegas per second.   d) arcs per second.

Answer: (b)
   MC   Easy

8.6: A phonograph record rotates at 45 rpm. Through what angle does it turn in 0.2 s?
   a) 9°   b) 15°   c) 54°   d) 96°

Answer: (c)
   MC   Moderate

8.7: A wheel starts at rest, and has an angular acceleration of 4 rad/s^2. Through what angle does it turn in 3 s?
   a) 36 rad   b) 18 rad   c) 12 rad   d) 9 rad

Answer: (b)
   MC   Easy

8.8: A boy and a girl are riding a merry-go-round which is turning at a constant rate. The boy is near the outer edge, while the girl is closer to the center. Who has the greater centripetal acceleration?
   a) The boy
   b) The girl
   c) Both have the same non-zero centripetal acceleration.
   d) Both have zero centripetal acceleration.

Answer: (a)
   MC   Easy

8.9: A non-zero net torque will
   a) cause a change in angular velocity.
   b) maintain a constant angular velocity.
   c) cause linear acceleration.
   d) maintain a constant angular momentum.

Answer: (a)
   MC   Easy

8.10: An ice skater is in a spin with his arms outstretched. If he pulls in his arms, what happens to his kinetic energy?
   a) It increases.
   b) It decreases.
   c) It remains constant but non-zero.
   d) It remains zero.

Answer: (a)
   MC   Moderate

8.11: The earth orbits the sun in an elliptical orbit. Ignore any friction which may be present. What happens over time to the earth's angular momentum about the sun?
   a) It continually increases.
   b) It continually decreases.
   c) It remains constant.
   d) It increases during some parts of the orbit, and decreases during others.

Answer: (c)
   MC   Moderate

8.12: A uniform solid sphere has mass M and radius R. If these are increased to 2M and 3R, what happens to the sphere's moment of inertia about a central axis?
   a) Increases by a factor of 6.
   b) Increases by a factor of 12.
   c) Increases by a factor of 18.
   d) Increases by a factor of 54.

Answer: (c)
   MC   Moderate

8.13: "The total angular momentum of a system of particles changes when a net external force acts on the system." This statement is
   a) always true.
   b) never true.
   c) sometimes true. It depends on the force's magnitude.
   d) sometimes true. It depends on the force's point of application.

Answer: (d)
   MC   Moderate

8.14: "The total angular momentum of a system of particles changes when a net external torque acts on the system." This statement is
   a) always true.
   b) never true.
   c) sometimes true. It depends on the torque's magnitude.
   d) sometimes true. It depends on the torque's point of application.

Answer: (a)
   MC   Moderate

8.15: In what circumstances can the angular velocity of system of particles change without any change in the system's angular momentum?
a) This cannot happen under any circumstances.
b) This can happen if a net external force acts on the system's center of mass.
c) This can happen if the only forces acting are internal to the system.
d) This can happen if an external net torque is applied properly to the system.

Answer: (c)
MC Moderate

8.16: What is the quantity used to measure an object's resistance to changes in rotation?
a) mass   b) moment of inertia
c) linear momentum   d) angular momentum

Answer: (b)
MC Easy

8.17: An ice skater has a moment of inertia of 5 kg·m^2 when her arms are outstretched. At this time she is spinning at 3 revolutions per second (rps). If she pulls in her arms and decreases her moment of inertia to 2 kg·m^2, how fast will she be spinning?
a) 2 rps   b) 3.3 rps   c) 7.5 rps   d) 10.3 rps

Answer: (c)
MC Easy

8.18: An ice skater is in a tight spin. He stretches out his arms and slows down. In doing so, angular momentum is conserved.

Answer: True
TF Easy

8.19: An ice skater is in a tight spin. He stretches out his arms and slows down. In doing so, linear momentum is conserved.

Answer: True
TF Easy

8.20: An ice skater is in a tight spin. He stretches out his arms and slows down. In doing so, kinetic energy is conserved.

Answer: False
TF Moderate

8.21: Angular momentum is the quantity used to measure an object's resistance to changes in rotation.

Answer: False
   TF   Easy

8.22: A wheel is rotating at 9.5 rpm. Through what angle will a point on the rim turn in 5 s?
   a) 57°   b) 285°   c) 360°   d) 3420°

Answer: (b)
   MC   Easy

8.23: A wheel of diameter 26 cm turns at 1500 rpm. How far will a point on the outer rim move in 2 s?
   a) 314 cm   b) 4084 cm   c) 8995.5 cm   d) 17990.8 cm

Answer: (b)
   MC   Moderate

8.24: A cable car at a ski resort carries skiers a distance of 6.8 km. The cable which moves the car is driven by a pulley with diameter 3 m. Assuming no slippage, how fast must the pulley rotate for the cable car to make the trip in 12 minutes?
   a) 9.4 rpm   b) 30.1 rpm   c) 60.1 rpm   d) 721.5 rpm

Answer: (c)
   MC   Difficult

8.1: Match the unit to the physical quantity.

8.25: angular accelaration

Answer: rad/s^2
   MA   Moderate

8.26: frequency

Answer: Hz
   MA   Easy

8.27: angular velocity

Answer: rad/s
   MA   Easy

8.28: period

Answer: seconds
   MA   Easy

8.29: angular displacement

Answer: radians
MA   Easy

8.30: What is the arc length subtended by an angle of 30° on a circle of radius 10 cm?

Answer: 5.24 cm
ES   Easy

8.31: A triangle has angles of 1.10 radians and 0.25 radians. What is the third angle, expressed in radians and in degrees?

Answer: 1.79 rad, and 103°
ES   Moderate

8.32: A bicycle wheel has an outside diameter of 66 cm. How far does a point on the rim travel when the wheel rotates through an angle of 70°?

Answer: 40.3 cm
ES   Moderate

8.33: A flywheel, of radius 1 m, is spinning with a constant angular velocity of 2 rad/s. What is the centripetal acceleration of a point on the wheel's rim?
a) $1 m/s^2$ b) $2 m/s^2$
c) $4 m/s^2$ d) None of the above

Answer: (c)
MC   Moderate

8.34: A tooth polisher on a dentist's drill reaches a frequency of 1800 rpm in 8 s. What is its average angular acceleration?

Answer: $23.6 rad/s^2$
ES   Easy

8.35: What is the angular speed of an electric motor that rotates at 1800 rpm?

Answer: 188 rad/s
ES   Easy

8.36: A fan blade, whose diameter is 1 m, is turning with an angular velocity of 2 rad/s. What is the tangential velocity of a point on the tip of the blade?
   a) 2 m/s
   b) 0.5 m/s
   c) 1 m/s
   d) None of the above

Answer: (c)
   MC   Easy

8.37: What is the angular speed of a record that rotates a 33-1/3 rpm?

Answer: 3.49 rad/s
   ES   Easy

8.38: The flywheel of a machine is rotating at 126 rad/s. Through what angle (in degrees) will the wheel be displaced from its original position after 6 s?

Answer: 116°
   ES   Moderate

8.39: What is the angular speed of a point on the earth's surface at 60° north latitude? What is the linear speed of such a point?

Answer: 7.27×10^{-5} rad/s; 0.232 km/s
   ES   Moderate

8.40: What is the centripetal acceleration of a point on the perimeter of a bicycle wheel of diameter 70 cm when the bike is moving 8 m/s?
   a) 91 m/s^2
   b) 183 m/s^2
   c) 206 m/s^2
   d) 266 m/s^2

Answer: (b)
   MC   Moderate

8.41: The diameter of the sun is 13.9×10^8 m, and the distance from the earth to the sun is 1.5×10^{11} m. In a sunset, how long does it take for the sun to disappear after it touches the horizon?

Answer: 2.12 min
   ES   Moderate

8.42: The rotating space station you're standing in simulates earth's gravity when you are 500 ft from its center. What must its angular speed be?
   a) 0.064 rad/s
   b) 0.14 rad/s
   c) 0.25 rad/s
   d) 126 rad/s

Answer: (b)
   MC   Easy

8.43: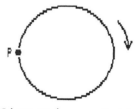

Marilyn (M) and her sister Sheila (S) are riding on a merry-go-round, as shown (bird's-eye view).
a) They have the same speed, but their angular velocity is different.
b) They have different speeds, and different angular velocities.
c) They have different speeds, but the same angular velocity.
d) They have the same speed and the same angular velocity.

Answer: (c)
MC   Easy

8.44: A pulsar (a rotating neutron star) emits pulses at a frequency of 0.4 kHz. The period of its rotation is
a) 2.5 ms   b) 2.5 s   c) 0.025 s   d) 25 ms

Answer: (a)
MC   Easy

8.45:

The record playing on the turntable of your stereo is rotating clockwise (as seen from above). After turning it off, your turntable is slowing down, but hasn't stopped yet. The direction of the acceleration of point P (at the left) is

a)

b)

c)

d)

e)

Answer: (a)
MC   Moderate

8.46: A future use of space stations may be to provide hospitals for severely burned persons. It is very painful for a badly burned person on earth to lie in bed. In a space station, the effect of gravity can be reduced or eliminated, At what frequency, in rpm, would a doughnut-shaped hospital of 200 m radius have to rotate, if persons on the outer perimeter are to experience 1/10 the gravity effect of earth?

Answer: 0.67 rpm
ES   Moderate

8.47: angular momentum

Answer: kg·m^2/s
MA   Easy

8.48: moment of inertia

Answer: kg·m^2
MA   Easy

8.49: torque

Answer: N·m
MA   Easy

8.50: rotational kinetic energy

Answer: joules
MA   Easy

8.51: lever-arm

Answer: meters
MA   Easy

8.52: A non-zero torque is needed to produce a change in angular velocity.

Answer: True
TF   Easy

8.53: Consider a rigid body that is rotating. Which of the following is an accurate statement?
   a) Its center of rotation is its center of gravity.
   b) All points on the body are moving with the same angular velocity.
   c) All points on the body are moving with the same linear velocity.
   d) Its center of rotation is at rest, i.e., not moving.

Answer: (b)
   MC   Moderate

8.54: A bicycle is moving 4 m/s. What is the angular speed of a wheel if its radius is 30 cm?
   a) 0.36 rad/s   b) 1.2 rad/s   c) 4.8 rad/s   d) 13.3 rad/s

Answer: (d)
   MC   Easy

8.55: A triatomic molecule is modeled as follows: mass m is at the origin, mass 2m is at x = a, and, mass 3m is at x = 2a. (a) What is the moment of inertia about the origin? (b) What is the moment of inertia about the center of mass?

Answer: (a) 14 ma^2 (b)   3.3 ma^2
   ES   Difficult

8.56: A cylinder, of radius 8 cm, rolls 20 cm in 5 s. Through what angular displacement (in degrees) does the cylinder move in this time?

Answer: 143°
   ES   Easy

8.57: A wheel of diameter 0.7 m rolls without slipping. A point at the top of the wheel moves with a tangential speed 2 m/s. (a) At what speed is the axis of the wheel moving? (b) What is the angular speed of the wheel?

Answer: (a) 1 m/s   (b)   2.86 rad/s
   ES   Moderate

8.58: Consider a bicycle wheel to be a ring of radius 30 cm and mass 1.5 kg. Neglect the mass of the axle and sprocket. If a force of 20 N is applied tangentially to a sprocket of radius 4 cm for 4 s, what linear speed does the wheel achieve, assuming it rolls without slipping?

Answer: 7.11 m/s
   ES   Difficult

8.59: A cylinder of radius R and mass M rolls without slipping down a plane inclined at an angle θ above horizontal. (a) What is the torque about the point of contact with the plane? (b) What is the moment of inertia about the center of the cylinder? (c) What is the linear speed of the cylinder one second after it is released from rest?

Answer: (a) MgR sin θ (b) $MR^2/2$ (c) 2gt sin θ
ES Difficult

8.60: A boy and a girl are riding on a merry-go-round that is turning. The boy is twice as far as the girl from the merry-go-round's center. If the boy and girl are of equal mass, which statement is true about the boy's moment of inertia with respect to the axis of rotation?
a) His moment of inertia is 4 times the girl's.
b) His moment of inertia is twice the girl's.
c) The moment of inertia is the same for both.
d) The boy has a greater moment of inertia, but it is impossible to say exactly how much more.

Answer: (a)
MC Moderate

8.61: Consider a motorcycle of mass 150 kg, one wheel of which has a mass of 10 kg and a radius of 30 cm. What is the ratio of the rotational kinetic energy of the wheels to the total translational kinetic energy of the bike? Assume the wheels are uniform disks.

Answer: 0.067:1
ES Difficult

8.62: A solid sphere of radius R and mass M is released from rest, and rolls without slipping down a plane inclined at angle θ above horizontal. What is the speed of the sphere after it has moved a distance L?

Answer: $(10/7\ gL\ sin\ \theta)^{1/2}$
ES Difficult

8.63: A hoop of radius 0.5 m and a mass of 0.2 kg is released from rest and allowed to roll down an inclined plane. How fast is it moving after dropping a vertical distance of 3 m?
a) 2   m/s   b) 3.8 m/s   c) 5.4 m/s   d) 7.7 m/s

Answer: (c)
MC Moderate

8.64: We can best understand how a diver is able to control his rate of rotation while in the air (and thus enter the water in a vertical position) by observing that while in the air,
a) his total energy is constant.
b) his kinetic energy is constant.
c) his potential energy is constant.
d) his angular momentum is constant.
e) his linear momentum is constant.

Answer: (d)
MC   Easy

8.65: A diver can change her rate of rotation in the air by "tucking" her head in and bending her knees. Let's assume that when she is stretched out straight she is rotating at 1 revolution per second. Now she goes into the "tuck and bend", effectively shortening the length of her body by half. What will her rate of rotation be now?

Answer: 4 Hz
ES   Moderate

8.66: A solid cylinder and a hollow cylinder have the same mass and the same radius. Which statement is true concerning their moment of inertia about an axis through the exact center of the flat surfaces?
a) The hollow cylinder has the greater moment of inertia.
b) The solid cylinder has the greater moment of inertia.
c) Both cylinders have the same moment of inertia.
d) The moment of inertia cannot be determined since it depends on the amount of material removed from the inside of the hollow cylinder.

Answer: (a)
MC   Easy

8.67: An ice skater performs a pirouette (a fast spin) by pulling in his outstretched arms close to his body. What happens to his moment of inertia about the axis of rotation?
a) It does not change.
b) It increases.
c) It decreases.
d) It changes, but it is impossible to tell which way.

Answer: (c)
MC   Easy

8.68: An ice skater performs a pirouette (a fast spin) by pulling in his outstretched arms close to his body. What happens to his angular momentum about the axis of rotation?
    a) It does not change.
    b) It increases.
    c) It decreases.
    d) It changes, but it is impossible to tell which way.

Answer: (a)
    MC   Easy

8.69: An ice skater performs a pirouette (a fast spin) by pulling in his outstretched arms close to his body. What happens to his rotational kinetic energy about the axis of rotation?
    a) It does not change.
    b) It increases.
    c) It decreases.
    d) It changes, but it is impossible to tell which way.

Answer: (b)
    MC   Moderate

8.70: Two uniform solid spheres have the same mass, but one has twice the radius of the other. The ratio of the larger sphere's moment of inertia to that of the smaller sphere is
    a) 4/5   b) 8/5   c) 2   d) 4

Answer: (d)
    MC   Moderate

8.71: Consider two uniform solid spheres where one has twice the mass and twice the diameter of the other. The ratio of the larger moment of inertia to that of the smaller moment of inertia is
    a) 2   b) 8   c) 4   d) 10   e) 6

Answer: (b)
    MC   Moderate

8.72: Consider two uniform solid spheres where both have the same diameter, but one has twice the mass of the other. The ratio of the larger moment of inertia to that of the smaller moment of inertia is
    a) 2   b) 8   c) 10   d) 4   e) 6

Answer: (a)
    MC   Moderate

8.73: An object's angular momentum changes by 10 kg-m^2/s in 2 s. What magnitude average torque acted on this object?
    a) 40 N-m   b) 2.5 N-m   c) 20 N-m   d) 5 N-m

Answer: (d)
    MC   Easy

8.74: The moment of inertia of a solid cylinder about its axis is given by $0.5MR^2$. If this cylinder rolls without slipping, the ratio of its rotational kinetic energy to its translational kinetic energy is

   a) 1:1   b) 1:2   c) 2:1   d) 1:3

Answer: (b)
   MC   Moderate

8.75: Consider two equal mass cylinders rolling with the same translational velocity. The first cylinder (radius = R) is hollow and has a moment of inertia about its rotational axis of MR^2, while the second cylinder (radius = r) is solid and has a moment of inertia about its axis of $0.5 Mr^2$. What is the ratio of the hollow cylinder's angular momentum to that of the solid cylinder?

   a) r^2 to $2R^2$ b) $2R^2$ to r^2 c) r to 2R   d) 2R to r

Answer: (d)
   MC   Difficult

8.76: A wheel of moment of inertia of 5.00 kg-m^2 starts from rest and accelerates under a constant torque of 3.00 N-m for 8.00 s. What is the wheel's rotational kinetic energy at the end of 8.00 s?

   a) 57.6 J   b) 64.0 J   c) 78.8 J   d) 122 J

Answer: (a)
   MC   Moderate

8.77: Two children, each of mass 20 kg, ride on the perimeter of a small merry-go-round that is rotating at the rate of 1 revolution every 4 s. The merry-go-round is a disk of mass 30 kg and radius 3 m. The children now both move halfway in toward the center to positions 1.5 m from the axis of rotation. Calculate the kinetic energy before and after, and the final rate of rotation.

Answer: K_O = 611 J, K = 1340 J, f = 0.55 Hz.
   ES   Difficult

8.78: A proton of mass m rotates with an angular speed of 2×10^6 rad/s in a circle of radius 0.8 m in a cyclotron. What is the orbital angular momentum of the proton?

   a) 1.28×10^6 m   b) 1.76×10^6 m
   c) 3.2×10^6 m   d) 6.4×10^6 m

Answer: (a)
   MC   Moderate

8.79: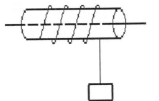

A uniform solid cylinder (mass 100 kg and radius 50 cm) is mounted so it is free to rotate about a fixed horizontal axis that passes through the centers of its circular ends. A 10-kg block is hung from a massless cord wrapped around the cylinder's circumference. When the block is released, the cord unwinds and the block accelerates downward. What is the block's acceleration?

Answer: 1.63 m/s^2
ES Difficult

8.80: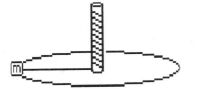

Initially, a 2.00-kg mass is whirling at the end of a string (in a circular path of radius 0.750 m) on a horizontal frictionless surface with a tangential speed of 5 m/s. The string has been slowly winding around a vertical rod, and a few seconds later the length of the string has shortened to 0.250 m. What is the instantaneous speed of the mass at the moment the string reaches a length of 0.250 m?
  a) 3.9 m/s   b) 15 m/s   c) 75 m/s   d) 225 m/s

Answer: (b)
MC Moderate

9. Bodies in Equilibrium; Elasticity and Fracture

9.1: What condition or conditions are necessary for rotational equilibrium?
   a) $\Sigma F_X = 0$
   b) $\Sigma F_X = 0$, $\Sigma F_Y = 0$, $\Sigma \tau = 0$
   c) $\Sigma \tau = 0$
   d) $\Sigma F_Y = 0$
   e) $\Sigma F_X = 0$, $\Sigma F_Y = 0$

Answer: (c)
   MC   Easy

9.2: What condition or conditions are necessary for static equilibrium?
   a) $\Sigma F_X = 0$
   b) $\Sigma F_X = 0$, $\Sigma F_Y = 0$, $\Sigma \tau = 0$
   c) $\Sigma \tau = 0$
   d) $\Sigma F_Y = 0$
   e) $\Sigma F_X = 0$, $\Sigma F_Y = 0$

Answer: (b)
   MC   Easy

9.3: A boy and a girl are balanced on a massless seesaw. The boy has a mass of 75 kg and the girl's mass is 50 kg. If the boy sits 2 m from the pivot point on one side of the seesaw, where must the girl sit on the other side?
   a) 1.33 m
   b) 2.5 m
   c) 3 m
   d) None of the above

Answer: (c)
   MC   Easy

9.4: A heavy boy and a lightweight girl are balanced on a massless seesaw. If they both move forward so that they are one-half their original distance from the pivot point, what will happen to the seesaw?
   a) The side the boy is sitting on will tilt downward.
   b) The side the girl is sitting on will tilt downward.
   c) Nothing, the seesaw will still be balanced.
   d) It is impossible to say without knowing the masses and the distances.

Answer: (c)
   MC   Moderate

9.5: A heavy seesaw is out of balance. A lightweight girl sits on the end that is tilted downward, and a heavy boy sits on the other side so that the seesaw now balances. If the boy and girl both move forward so that they are one-half their original distance from the pivot point, what will happen to the seesaw?
 a) The side the boy is sitting on will now tilt downward.
 b) The side the girl is sitting on will once again tilt downward.
 c) Nothing, the seesaw will still be balanced.
 d) It is impossible to say without knowing the masses and the distances.

Answer: (b)
 MC   Moderate

9.6: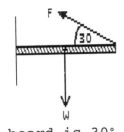
A force is applied to the end of a 2 m long uniform board weighing 50 N, in order to keep it horizontal, while it pushes against a wall at the left. If the angle the force makes with the board is 30° in the direction shown, the applied force F is
 a) 28.9 N   b) 50 N   c) 57.7 N   d) 100 N

Answer: (b)
 MC   Moderate

9.7: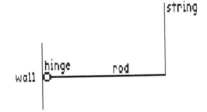
A uniform rod has a weight of 40 N and a length of 1 m. It is hinged to a wall (at the left end), and held in a horizontal position by a vertical massless string (at the right end). What is the magnitude of the torque exerted by the string about a horizontal axis which passes through the hinge and is perpendicular to the rod?
 a) 5 N-m   b) 10 N-m   c) 20 N-m   d) 40 N-m

Answer: (c)
 MC   Moderate

9.8: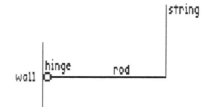

A uniform rod has a length of 2 m. It is hinged to a wall (at the left end), and held in a horizontal position by a vertical massless string (at the right end). What is the angular acceleration of the rod at the moment the string is released?

a) 3.27 rad/s^2
b) 14.7 rad/s^2
c) 7.35 rad/s^2
d) It can't be calculated without knowing the rod's mass.

Answer: (c)
MC Difficult

9.9:

Five forces act on a massless rod free to pivot at point P. Which force is producing a counter-clockwise torque about point P?

a) A   b) B   c) C   d) D   e) E

Answer: (c)
MC Easy

9.10: Stress is
a) the strain per unit length.
b) the same as force.
c) the ratio of the change in length.
d) applied force per cross-sectional area.

Answer: (d)
MC Easy

9.11: Strain is
a) the ratio of the change in length to the original length.
b) the stress per unit area.
c) the applied force per unit area.
d) the ratio of stress to elastic modulus.

Answer: (a)
MC Easy

9.12:

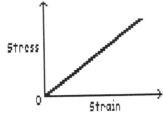

The slope of the straight line shown on the graph is called the
a) bulk modulus.
b) Young's modulus.
c) compressibility.
d) pressure.

Answer: (b)
MC Moderate

9.13: A mass of 50 kg is suspended from a steel wire of diameter 1 mm and length 11.2 m. How much will the wire stretch? The Young's modulus for steel is 20 x 10^10 N/m^2.

Answer: 3.5 cm
ES Moderate

9.14: At a depth of about 1030 m in the sea the pressure has increased by 100 atmospheres (to about 10^7 N/m^2). By how much has 1 m^3 of water been compressed by this pressure? The bulk modulus of water is 2.3 x 10^9 N/m^2.

Answer: 0.0043 m^3
ES Moderate

9.15: The weight that will cause a wire of diameter d to stretch a given distance, for a fixed length of wire, is
a) independent of what the length of the wire is.
b) proportional to d.
c) proportional to d^2.
d) independent of d.

Answer: (c)
MC Easy

9.16: Suppose that an 80-kg person walking on crutches supports all his weight on the two crutch tips, each of which is circular with a diameter of 4 cm. What pressure is exerted on the floor? Express the answer in psi.
a) 78 kPa   b) 156 kPa   c) 312 kPa   d) 624 kPa

Answer: (c)
MC Moderate

9.17: A copper wire is found to break when subjected to minimum tension of 36 N. If the wire diameter were half as great, we would expect the wire to break when subjected to a minimum tension of
a) 9 N   b) 18 N   c) 25.5 N   d) 36 N   e) 50.7 N

Answer: (a)
MC   Easy

9.18: A rocket moves through outer space with a constant velocity of 9.8 m/s. What net force acts on it?
a) A force equal to its weight on earth, mg.
b) A force equal to the gravity acting on it.
c) The net force is zero.
d) Cannot be determined without more information.

Answer: (c)
MC   Easy

9.19: A book weighs 6 N. When held at rest above your head the net force on the book is
a) 0 N   b) 6 N   c) 9.8 N   d) -6 N

Answer: (a)
MC   Easy

9.20: A person weighing 800 N stands with one foot on each of two bathroom scales. Which statement is definitely true?
a) Each scale will read 800 N.
b) Each scale will read 400 N.
c) If one scale reads 500 N, the other will read 300 N.
d) None of the above is definitely true.

Answer: (c)
MC   Moderate

9.21: Two telephone poles are separated by 40 m and connected by a wire. A bird of mass 0.5 kg lands on the wire midway between the poles, causing the wire to sag 2m below horizontal. Assuming the wire has negligible mass, what is the tension in the wire?

Answer: 24.6 N
ES   Difficult

9.22: Consider two identical bricks, each of dimensions 20 cm x 10 cm x 6 cm. One is stacked on the other, and the combination is then placed so that they project out over the edge of a table. What is the maximum distance that the end of the top brick can extend beyond the table edge without toppling?
a) 7.5 cm   b) 10 cm   c) 12.5 cm   d) 15 cm

Answer: (d)
MC   Moderate

9.23: Two scales are separated by 2 m, and a plank of mass 4 kg is placed between them. Each scale is observed to read 2 kg. A person now lies on the plank, after which the right scale reads 30 kg and the left scale reads 50 kg. How far from the right scale is the person's center of gravity located?

    a) 1.20 m   b) 1.25 m   c) 1.26 m   d) 1.32 m

Answer: (c)
    MC   Moderate

9.24: Consider a hydraulic press. When the input piston is depressed 16 cm the output piston moves 1 cm. Thus in the absence of friction we expect that a force of 16 N applied to the input piston will result in a force exerted by the output piston of

    a) 1 N   b) 16 N
    c) 256 N   d) more than 256 N

Answer: (c)
    MC   Easy

9.25: A worker uses a lever to lift a crate. The crate weighs 400 N, and the worker exerts 100 N to lift it. The distance from crate to fulcrum is 24 cm, and the distance from worker to fulcrum is 108 cm. What are the actual mechanical advantage and the theoretical mechanical advantage?

Answer: AMA = 4.0
    TMA = 4.5
    ES   Moderate

9.26: A machine with an efficiency of 75% and input force of 600 N moves through a parallel distance of 0.4 m. How much energy is lost in this process?

    a) 60 J   b) 80 J   c) 180 J   d) 240 J

Answer: (a)
    MC   Moderate

9.27: A machine with an AMA of 6 loses 8% of the input work to friction. What is the TMA of the machine?

    a) 5.52   b) 6.52   c) 6.84   d) 7.02

Answer: (b)
    MC   Easy

9.28: The theoretical mechanical advantage of a block and tackle with N support strands is

    a) N - 1   b) N   c) N + 1   d) (2N + 1)/2

Answer: (b)
    MC   Easy

9.29: What is a "machine"?
   a) A device that provides energy to do work.
   b) A device used to change the magnitude or direction of a force.
   c) A device to convert energy into momentum, or vice versa.
   d) A device which always changes a small force into a larger one.

Answer: (b)
   MC   Easy

9.30: The base of a packing crate measures 1.2 m x 1.2 m, and it is 1.6 m tall. It weighs 400 N. Assuming it does not slip on the floor, what is the minimum force one can apply in order to tip the crate?
   a) 80 N   b) 120 N   c) 200 N   d) 400 N

Answer: (b)
   MC   Moderate

9.31: A passenger van has an outer wheel base width of 2 m. Its center of gravity is equidistant from the sides, and positioned 1.2 m above the ground. What is the maximum sideways angle at which it can be inclined without tipping over?
   a) 32.6°   b) 39.8°   c) 50.2°   d) 57.3°

Answer: (b)
   MC   Moderate

9.32: Two children sit on opposite ends of a uniform seesaw which pivots in the center. Child A has mass 60 kg and sits 2 m from the center. Child B has mass 40 kg. How far from the center must child B sit for the seesaw to balance?
   a) 1.33 m
   b) 2.5 m
   c) 3 m
   d) Cannot be determined without knowing the seesaw's mass.

Answer: (c)
   MC   Easy

9.33: A machine may not be used to
   a) multiply available force.
   b) create additional energy.
   c) transfer energy from one place to another.
   d) change the direction of an applied force.

Answer: (b)
   MC   Easy

9.34: A mechanic lifts the front end of a car using a jack. He exerts
a 120 N force on the jack handle. The front end of the car
weighs 4800 N. Which is true?
    a) TMA = 40   b) AMA = 40
    c) TMA = 4800 N   d) AMA = 120

Answer: (b)
    MC   Easy

9.35: A lever is 5 m long. The distance from the fulcrum to the weight
to be lifted is 1 m. If a worker pushes on the opposite end with
400 N, what is the maximum weight that can be lifted?
    a) 80 N   b) 100 N   c) 1600 N   d) 2000 N

Answer: (c)
    MC   Easy

9.36: A lever is 5 m long. The distance from the fulcrum to the weight
to be lifted is 1 m. If a 3000 N rock is to be lifted, how much
force must be exerted on the lever?
    a) 600 N   b) 750 N   c) 3000 N   d) 12000 N

Answer: (b)
    MC   Easy

9.37: A massless scaffold is held up by a wire at each end. The
scaffold is 12 m long. A 300 N box sits 3 m from the left end.
What is the tension in each wire?
    a) Left wire = 100 N; right wire = 200 N
    b) Left wire = 200 N; right wire = 100 N
    c) Left wire = 900 N; right wire = 2700 N
    d) Left wire = 2700 N; right wire = 900 N

Answer: (b)
    MC   Moderate

9.38: A 200 N scaffold is held up by a wire at each end. The scaffold
is 18 m long. A 650 N box sits 3 m from the left end. What is
the tension in each wire?
    a) Left wire = 520 N; right wire = 130 N
    b) Left wire = 620 N; right wire = 230 N
    c) Left wire = 195 N; right wire = 975 N
    d) Left wire = 295 N; right wire = 1000 N

Answer: (b)
    MC   Moderate

9.39: A 100 kg scaffold is 6 m long. It is hanging from wires on each
end. A 500 kg box sits 1 m from the left end. What is the
tension in each wire?
a) Left wire = 417 kg; right wire = 83 kg
b) Left wire = 467 N;   right wire = 133 N
c) Left wire = 4083 N; right wire = 817 N
d) Left wire = 4573 N; right wire = 1307 N

Answer: (d)
MC Moderate

9.40: A 5 m long scaffold weighs 80 N. It is hanging from wires at
each end. A 200 N box sits 1 m from the left end. What is the
tension in each wire?
a) Left wire = 40 N; right wire = 160 N
b) left wire = 200 N; right wire = 80 N
c) Left wire = 160 N; right wire = 40 N
d) Left wire = 240 N; right wire = 120 N

Answer: (b)
MC Moderate

9.41: A bridge piling has an area of 1.25 m^2. It supports 1875 N.
Find the stress on the column.
a) 1875 N   b) 1875 N/m^2 c) 1500 N/m^2 d) 2344 N/m^2

Answer: (c)
MC Easy

9.42: A 5000 N force compresses a steel block by 0.0025 cm. How much
force would be needed to compress the block by 0.0125 cm?
a) 1000 N   b) 2500 N   c) 5000 N   d) 25000 N

Answer: (d)
MC Moderate

9.43: A cable is 100 m long and has a cross-section area of 1 mm^2. A
1000 N force is applied to stretch the cable. The elastic
modulus for the cable is 1 x 10^{11} N/m^2. How far does it stretch?
a) 0.01 m   b) 0.10 m   c) 1.0 m   d) 10 m

Answer: (c)
MC Moderate

9.44: A wire of diameter 0.2 mm stretches by 0.2% when a 6.28 N force
is applied. What is the elastic modulus of the wire?
a) 2.5 x 10^{10} N/m^2 b) 1.0 x 10^{11} N/m^2
c) 2.5 x 10^{12} N/m^2 d) 1.0 x 10^{12} N/m^2

Answer: (b)
MC Moderate

9.45: A steel lift column in a service station is 4m long and 0.2 m in diameter. Young's modulus for steel is 20×10^{10} N/m^2. By how much does the column shrink when a 5000 kg truck is on it?
    a) 8.0×10^{-7} m   b) 3.2×10^{-6} m
    c) 7.8×10^{-6} m   d) 3.1×10^{-5} m

Answer: (d)
    MC   Moderate

10. Fluids

10.1: Match the unit to the physical quantity.

10.1: flow rate

Answer: m^3/s
MA   Easy

10.2: Reynolds number

Answer: dimensionless
MA   Easy

10.3: surface tension

Answer: N/m
MA   Easy

10.4: 1 mm Hg

Answer: torr
MA   Easy

10.5: 1 atmosphere

Answer: $14.7 \ lb/in^2$
MA   Easy

10.6: SI unit for viscosity

Answer: Pa-s
MA   Easy

10.7: What is the increased pressure on a submarine at a depth of 100 m, compared to the surface?

Answer: 9.8×10^5 Pa
ES   Easy

10.8: A window for viewing fish in the local aquarium is square and measures 0.8 m x 0.8 m. Its center is 12 m below the surface. What is the total force on the window?

Answer: 75,300 N
ES   Moderate

10.9: Suppose that in an examination of a spinal injury it is found that the spinal fluid (with about the same density as water) will rise a vertical height of 13 cm in an open tube. What pressure does this correspond to, expressed in mm of mercury and in atmospheres?

Answer: 9.56 mm Hg or 0.013 atm
ES   Moderate

10.10: A plastic block of dimensions 2 cm x 3 cm x 4 cm has a mass of 30 g. What is its density?
   a) 0.8 g/cm^3 b) 1.2 g/cm^3 c) 1.25 g/cm^3 d) 1.60 g/cm^3

Answer: (c)
MC   Easy

10.11: A circular window of 30 cm diameter in a submarine can withstand a maximum force of 5.20 x 10^5 N. What is the maximum depth to which the submarine can go without damaging the window?
   a) 680 m   b) 750 m   c) 1200 m   d) 1327 m

Answer: (b)
MC   Moderate

10.12: Substance A has a density of 3 g/cm^3 and substance B has a density of 4 g/cm^3. In order to obtain equal masses of these two substances, the ratio of the volume of A to the volume of B will be equal to
   a) 1:3   b) 4:3   c) 3:4   d) 1:4

Answer: (b)
MC   Easy

10.13: A piece of iron sinks to the bottom of a lake where the pressure is 21 atm. Which statement best describes what happens to the volume of that piece of iron?
   a) Its volume decreases slightly.
   b) Its volume becomes 21 times greater.
   c) Its volume increases slightly.
   d) There has been no change in the volume of the iron.

Answer: (a)
MC   Easy

10.14: A piece of iron sinks to the bottom of a lake where the pressure is 21 atm. Which statement best describes what happens to the density of that piece of iron?
   a) Its density decreases slightly.
   b) Its density becomes 21 times greater.
   c) Its density increases slightly.
   d) There has been no change in the density of the iron.

Answer: (c)
MC   Moderate

10.15: A cylindrical rod of length 12 cm and diameter 2 cm will just barely float in water. What is its mass?

Answer: 37.7 g
ES   Moderate

10.16: A crane lifts a steel submarine (density = 7.8×10^3 kg/m^3) of mass 20,000 kg. What is the tension in the lifting cable (a) when the submarine is submerged, and (b) when it is entirely out of the water?

Answer: (a) 1.71×10^5 N;   (b) 1.96×10^5 N
ES   Moderate

10.17: An object completely submerged in water must either rise or fall.

Answer: False
TF   Easy

10.18: A polar bear of mass 200 kg stands on an ice floe 100 cm thick. What is the minimum area of the floe that will just support the bear in saltwater of specific gravity 1.03? The specific gravity of ice is 0.98.

Answer: 4.0 m^3
ES   Difficult

10.19: Oil of specific gravity 0.90 is poured on top of a beaker of water. A cube of plastic of specific gravity 0.95 is placed in the liquid. What fraction of the cube will be immersed in water?

Answer: 0.5
ES   Moderate

10.20: A sunken steel ship has a mass of 500,000 kg. It is filled with water. In order to lift the ship, air bags are to be inflated inside the hull. What volume of air is needed? The specific gravity of steel is 7.8.
    a) 225 m^3 b) 436 m^3 c) 1266 m^3 d) 2778 m^3

Answer: (b)
MC   Difficult

10.21: A container of water is placed on a scale, and the scale reads 120 g. Now a 20-g piece of copper (specific gravity = 8.9) is suspended from a thread and lowered into the water, not touching the bottom of the container. What will the scale now read?
    a) 120 g   b) 122 g   c) 138 g   d) 140 g

Answer: (b)
MC   Difficult

10.22: When a glass tube is inserted into mercury (Hg), the mercury level inside the tube is lower than the mercury level outside the tube.

Answer: True
TF Moderate

10.23: The surface tension of water is 0.073 N/m. How high will water rise in a capillary tube of diameter 1.2 mm?

Answer: 2.48 cm
ES Moderate

10.24: When a tube of diameter d is placed in water, the water rises to a height h. If the diameter were half as great, how high would the water rise?
   a) h/2   b) h   c) 2h   d) 4h

Answer: (c)
MC Easy

10.25: Consider a rectangular frame of length 0.120 m and width 0.0600 m with a soap film formed within its confined area. If the surface tension of soapy water is 0.0260 N/m, how much force does the soap film exert on the 0.600 m side?
   a) 0.00624 N   b) 0.00156 N
   c) 0.00312 N   d) None of the above.

Answer: (c)
MC Moderate

10.26: A narrow tube is placed vertically in a pan of water, and the water rises in the tube to 4 cm above the level of the pan. The surface tension in the liquid is lowered to one-half its original value by the addition of some soap. What happens to the height of the liquid column in the tube?
   a) It drops to 2 cm.
   b) It drops to 1 cm.
   c) It drops or rises, but the height cannot be calculated from the information given.
   d) It remains at the same height.

Answer: (a)
MC Moderate

10.27: Falling drops of milk tend to form spheres, in order to maximize their surface area.

Answer: True
TF Moderate

10.28: When soup gets cold, it often tastes greasy. This "greasy" taste
seems to be associated with oil spreading out all over the
surface of the soup, instead of staying in little globules. To
us "physikers", this is readily explained in terms of
a) the Bernoulli effect.
b) Archimedes Principle.
c) the decrease in the surface tension of water with increasing
temperature.
d) the increase in the surface tension of water with increasing
temperature.

Answer: (d)
MC   Moderate

10.29: A hole of radius 1 mm occurs in the bottom of a water storage
tank that holds water at a depth of 15 m. At what rate will
water flow out of the hole?

Answer: 0.054 L/s
ES   Difficult

10.30: Water flows through a horizontal pipe of cross-sectional area 10
cm^2 at a pressure of 0.25 atm. The flow rate is 1 L/s. At a
valve, the effective cross-sectional area of the pipe is reduced
to 5 cm^2. What is the pressure at the valve?
 a) 0.112 atm   b) 0.157 atm   c) 0.200 atm   d) 0.235 atm

Answer: (d)
MC   Difficult

10.31: Suppose that the build-up of fatty tissue on the wall of an
artery decreased the radius by 10%. By how much would the
pressure provided by the heart have to be increased to maintain a
constant blood flow?

Answer: 52%
ES   Moderate

10.32: SAE No. 10 oil has a viscosity of 0.2 Pa-s. How long would it
take to pour 4 L of oil through a funnel with a neck 15 cm long
and 2 cm in diameter? Assume the surface of the oil is kept 6 cm
above the top of the neck, and neglect any drag effects due to
the upper part of the funnel.
 a) 46 s   b) 52 s   c) 84 s   d) 105 s

Answer: (b)
MC   Difficult

10.33:

Air is forced to flow over an object whose cross-section is shown. The speed of the air flowing close to the top surface is greater than the speed of the air flowing close to the bottom surface.

Answer: False
TF   Easy

10.34: When a small spherical rock of radius r falls through water, it experiences a drag force (a)(r)(v), where "v" is its velocity and "a" is a constant proportional to the viscosity of water. From this, one can deduce that if a rock of diameter 2 mm falls with terminal velocity, "v", then a rock of diameter 4 mm will fall with terminal velocity

   a) v   b) $\sqrt{2v}$ c) 2 v   d) 4 v

Answer: (d)
MC   Difficult

10.35: Two styrofoam balls, of radii R and 2R, are released simultaneously from a tall tower. Which will reach the ground first?
   a) Both will reach the ground simultaneously.
   b) The larger one
   c) The smaller one
   d) The result will depend on the atmospheric pressure.

Answer: (b)
MC   Moderate

10.36: Which has the greatest effect on the flow of fluid through a narrow pipe? That is, if you made a 10% change in each of the quantities below, which would cause the greatest change in the flow rate?
   a) The fluid viscosity   b) The pressure difference
   c) The length of the pipe   d) The radius of the pipe

Answer: (d)
MC   Moderate

10.37: An adjustable rectangular frame has length 12 cm and width 5 cm, and is built in such a way that one of the longer sides can be moved without bursting the soap film contained in the frame. Soapy water has surface tension 0.026 N/m. How much work is required to move the adjustable long side so that the frame is 12 cm square?
   a) 0.0002184 J   b) 0.0004368 J
   c) 2.184 J   d) 4.368 J

Answer: (b)
MC   Moderate

10.38: Two narrow tubes are placed in a pan of water. Tube A has twice the diameter of tube B. If water rises 24 cm in tude A, how high will it rise in tube B?
a) 12 cm   b) 24 cm   c) 48 cm   d) 96 cm

Answer: (c)
MC   Easy

10.39: Certain insects, such as the water bug, are sufficiently lightweight that they can run on top of water without breaking the surface tension. Water bug A has weight W. Water bug B is twice as big as bug A, in all dimensions. That is, bug B is twice as long, twice as wide, etc. What is the weight of bug B?
a) 1.26 W   b) 2 W   c) 4 W   d) 8 W

Answer: (d)
MC   Moderate

10.40: Certains insects, such as the water bug, are sufficently lightweight that they can run on top of water without breaking the surface tension. This is possible because the water, due to surface tension, exerts an upward force on the bottom of the bug's feet. Suppose that the maximum possible upward force on the feet of water bug A is F. Now suppose that water bug B is twice as big as bug A in every dimension. That is, bug B is twice as long, twice as wide, etc. What is the maximum upward force on the feet of bug B?
a) 1.4 F   b) 2 F   c) 4 F   d) 8 F

Answer: (c)
MC   Moderate

10.41: Aluminum has bulk modulus 7.5×10^{10} N/m^2. A solid aluminum sphere has volume 2.00 m^3, and is subjected to an increased pressure of 2×10^7 Pa. What is the volume change of the sphere?

Answer: 5.3×10^{-4} m^3
ES   Moderate

10.42: Consider three drinking glasses. All three have the same area base, and all three are filled to the same depth with water. Glass A is cylindrical. Glass B is wider at the top than at the bottom, and so holds more water than A. Glass C is narrower at the top than at the bottom, and so holds less water than A. Which glass has the greatest liquid pressure at the bottom?
a) Glass A.
b) Glass B.
c) Glass C.
d) All three have equal pressure.

Answer: (d)
MC   Moderate

10.43: You are originally 1 m beneath the surface of a pool. If you dive to 2 m beneath the surface, what happens to the absolute pressure on you?
   a) It quadruples.
   b) It more than doubles.
   c) It doubles.
   d) It less than doubles.

Answer: (d)
   MC   Moderate

10.44: You are originally 1 m beneath the surface of a pool. If you dive to 3m, what happens to the additional pressure (in excess of atmospheric pressure) on you?
   a) Increases by a factor of nine.
   b) More than triples.
   c) Triples.
   d) Less than triples.

Answer: (c)
   MC   Easy

10.45: When atmospheric pressure changes, what happens to the absolute pressure at the bottom of a pool?
   a) It does not change.
   b) It increases by a lesser amnount.
   c) It increases by the same amount.
   d) It increases by a greater amount.

Answer: (c)
   MC   Moderate

10.46: A 500 N weight sits on the small piston of a hydraulic machine. The small piston has area 2 cm^2. If the large piston has area 40 cm^2, how much weight can the large piston support?
   a) 25 N   b) 500 N   c) 10000 N   d) 40000 N

Answer: (c)
   MC   Easy

10.47: As a rock sinks deeper and deeper into water of constant density, what happens to the buoyant force on it?
   a) It increases.
   b) It remains constant.
   c) It decreases.
   d) It may increase or decrease, depending on the shape of the rock.

Answer: (b)
   MC   Moderate

10.48: 50 cm^3 of wood is floating on water, and 50 cm^3 of iron is totally submerged. Which has the greater buoyant force on it?
   a) The wood.
   b) The iron.
   c) Both have the same buoyant force.
   d) Cannot be determined without knowing their densities.

Answer: (b)
   MC   Moderate

10.49: A 10 kg piece of aluminum sits at the bottom of a lake, right next to a 10 kg piece of lead. Which has the greater buoyant force on it?
   a) The aluminum.
   b) The lead.
   c) Both have the same buoyant force.
   d) Cannot be determined without knowing their volumes.

Answer: (a)
   MC   Moderate

10.50: Salt water has greater density than fresh water. A boat floats in both fresh water and in salt water. Where is the buoyant force greater on the boat?
   a) Salt water.
   b) Fresh water.
   c) Buoyant force is the same in both.
   d) Impossible to determine from the information given.

Answer: (c)
   MC   Moderate

10.51: A piece of iron rests on top of a piece of wood floating in a bathtub. If the iron is removed from the wood, what happens to the water level in the tub?
   a) It goes up.
   b) It goes down.
   c) It does not change.
   d) Impossible to determine from the information given.

Answer: (b)
   MC   Moderate

10.52: A piece of wood is floating in a bathtub. A second piece of wood sits on top of the first piece, and does not touch the water. If the top piece is taken off and placed in the water, what happens to the water level in the tub?
   a) It goes up.
   b) It goes down.
   c) It does not change.
   d) Cannot be determined from the information given.

Answer: (c)
   MC   Difficult

10.53: Salt water is more dense than fresh water. A ship floats in both fresh water and salt water. Compared to the fresh water, the amount of water displaced in the salt water is
a) more.
b) less.
c) the same.
d) Cannot be determined from the information given.

Answer: (b)
MC   Moderate

10.54: A steel ball sinks in water but floats in a pool of mercury. Where is the buoyant force on the ball greater?
a) Floating on the mercury.
b) Submerged in the water.
c) It is the same in both cases.
d) Cannot be determined from the information given.

Answer: (a)
MC   Moderate

10.55: An object floats with half its volume beneath the surface of the water. The weight of the displaced water is 2000 N. What is the weight of the object?
a) 1000 N
b) 2000 N
c) 4000 N
d) Cannot be determined from the information given.

Answer: (b)
MC   Moderate

10.56: A 200 N object floats with three-fourths of its volume beneath the surface of the water. What is the buoyant force on the object?
a) 50 N   b) 150 N   c) 200 N   d) 267 N

Answer: (c)
MC   Moderate

10.57: A solid object floats in water with three-fourths of its volume beneath the surface. What is the object's density?
a) 1333 kg/m^3 b) 1000 kg/m^3 c) 750 kg/m^3 d) 250 kg/m^3

Answer: (c)
MC   Moderate

10.58: A liquid has a specific gravity of 0.357. What is its density?
a) 357 kg/m^3 b) 643 kg/m^3 c) 1000 kg/m^3 d) 3570 kg/m^3

Answer: (a)
MC   Easy

10.59: As the speed of a moving fluid increases, the pressure in the fluid
   a) increases.
   b) remains constant.
   c) decreases.
   d) may increase or decrease, depending on the viscosity.

Answer: (c)
   MC   Easy

10.60: Two horizontal pipes are the same length, but pipe B has twice the diameter of pipe A. Water undergoes viscous flow in both pipes, subject to the same pressure difference across the lengths of the pipes. If the flow rate in pipe A is F, what is the flow rate in pipe B?
   a) 2 F   b) 4 F   c) 8 F   d) 16 F

Answer: (d)
   MC   Moderate

10.61: Two horizontal pipes have the same diameter, but pipe B is twice as long as pipe A. Water undergoes viscous flow in both pipes, subject to the same pressure difference across the lengths of the pipes. If the flow rate in pipe B is F, what is the flow rate in pipe A?
   a) F   b) 2 F   c) 4 F   d) 8 F

Answer: (b)
   MC   Moderate

10.62: A sky diver falls through the air at terminal velocity. The force of air resistance on him is
   a) half his weight.
   b) equal to his weight.
   c) twice his weight.
   d) Cannot be determined from the information given.

Answer: (b)
   MC   Easy

10.63: What is the difference between the pressures inside and outside a tire called?
   a) absolute pressure   b) atmospheric pressure
   c) gauge pressure   d) N/m^2

Answer: (c)
   MC   Easy

10.64: Which of the following is not a unit of pressure?
   a) atmosphere   b) N/m
   c) Pascal   d) mm of mercury

Answer: (b)
   MC   Easy

10.65: If outside air pressure changes, what happens to the reading on a tire gauge which has stayed connected to the tire?
   a) It increases.
   b) It remains constant.
   c) It decreases.
   d) It depends on the temperature.

Answer: (a)
   MC   Moderate

10.66: Water flows through a pipe. The diameter of the pipe at point B is larger than at point A. Where is the speed of the water greater?
   a) Point A
   b) Point B
   c) Same at both A and B
   d) Cannot be determined from the information given.

Answer: (a)
   MC   Easy

10.67: Water flows through a pipe. The diameter of the pipe at point B is larger than at point A. Where is the water pressure greatest?
   a) Point A
   b) Point B
   c) Same at both A and B
   d) Cannot be determined from the information given.

Answer: (b)
   MC   Moderate

10.68: A person weighing 900 N is standing on snowshoes. Each snowshoe has area 2500 cm^2. What is the pressure on the snow?
   a) 0.18 N/m^2 b) 0.36 N/m^2 c) 1800 N/m^2 d) 3600 N/m^2

Answer: (c)
   MC   Moderate

10.69: The upward force exerted on an object by a fluid is called the buoyant pressure.

Answer: False
   TF   Easy

10.70: The density of a substance divided by the density of water is called specific gravity.

Answer: True
   TF   Easy

10.71: In a flowing fluid, the pressure is highest where the fluid is flowing fastest.

Answer: False
TF   Easy

10.72: A brick weighs 50 N, and measures 30 cm x 10 cm x 4 cm. What is the maximum pressure it can exert on a horizontal surface?
a) 1.25 Pa   b) 12.5 Pa   c) 1.25 kPa   d) 12.5 kPa

Answer: (d)
MC   Easy

10.73: How much pressure does it take for a pump to supply a drinking fountain with 300 kPa, if the fountain is 30 m above the pump?
a) 294 kPa   b) 300 kPa   c) 594 kPa   d) 675 kPa

Answer: (c)
MC   Moderate

10.74: How much pressure must a submarine withstand at a depth of 120 m?
a) 123 N/m^2 b) 1205 N/m^2 c) 123 kPa   d) 1205 kPa

Answer: (d)
MC   Easy

10.75: A 13000 N vehicle is to be lifted by a 25 cm hydraulic piston. What force needs to be applied to a 5 cm diameter piston to accomplish this?
a) 260 N   b) 520 N   c) 2600 N   d) 5200 N

Answer: (b)
MC   Moderate

10.76: What is the gauge pressure if the absolute pressure is 300 kPa?
a) 97 kPa   b) 101 kPa   c) 199 kPa   d) 300 kPa

Answer: (c)
MC   Easy

10.77: A block of metal weighs 45 N in air and 25 N in water. What is the buoyant force of the water?
a) 20 N   b) 25 N   c) 45 N   d) 70 N

Answer: (a)
MC   Easy

10.78: An object has a volume of 4 m³ and weighs 40000 N. What will its weight be in water?
   a) 40000 N   b) 39200 N   c) 9800 N   d) 800 N

Answer: (d)
   MC   Difficult

10.79: A 1 m³ object floats in water with 20% of it above the waterline. What does the object weigh out of the water?
   a) 1960 N   b) 7480 N   c) 9800 N   d) 11760 N

Answer: (b)
   MC   Difficult

10.80: Liquid flows through a pipe of diameter 5 cm at 1.0 m/s. Find the flow rate.
   a) 0.00196 m³/s   b) 0.00785 m³/s
   c) 19.6 m³/s   d) 78.5 m³/s

Answer: (a)
   MC   Moderate

10.81: Liquid flows through a 4 cm diameter pipe at 1.0 m/s. There is a 2 cm diameter restriction in the line. What is the velocity in this restriction?
   a) 0.25 m/s   b) 0.50 m/s   c) 2 m/s   d) 4 m/s

Answer: (d)
   MC   Moderate

11. Vibrations and Waves

11.1: Match the unit to the physical quantity.

11.1: spring constant

Answer: N/m
       MA   Easy

11.2: linear mass density

Answer: g/cm
       MA   Easy

11.3: period

Answer: seconds
       MA   Easy

11.4: frequency

Answer: Hertz
       MA   Easy

11.5: wavelength

Answer: meters
       MA   Easy

11.2: The following question(s) refers to the wave shown below.

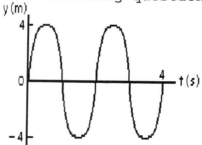

11.6: The amplitude is
       a) 2 m.   b) 4 m.   c) 6 m.   d) 8 m.

Answer: (b)
       MC   Easy   Table: 2

11.7: The wavelength
   a) is 2 m.
   b) is 4 m.
   c) is 8 m.
   d) cannot be determined from the given information.

Answer: (d)
   MC   Moderate   Table: 2

11.8: The frequency is
   a) 0.5 Hz.   b) 1 Hz.   c) 2 Hz.   d) 4 Hz.

Answer: (a)
   MC   Easy   Table: 2

11.9: A pendulum makes 12 complete swings in 8 s. (a) What is its
   frequency? (b) What is its period?

Answer: (a) 1.5 Hz   (b) 0.67 s
   ES   Easy

11.10: A mountain climber of mass 60 kg slips and falls a distance of 4
   m, at which time he reaches the end of his elastic safety rope.
   The rope then stretches an additional 2 m before the climber
   comes to rest. What is the spring constant of the rope, assuming
   it obeys Hooke's law?

Answer: 1760 N/m
   ES   Moderate

11.11: What is the spring constant of a spring that stretches 2 cm when
   a mass of 0.6 kg is suspended from it?

Answer: 294 N/m
   ES   Easy

11.12: What happens to a simple pendulum's frequency if both its length
   and mass are increased?
   a) It increases.
   b) It decreases.
   c) It remains constant.
   d) It could remain constant, increase, or decrease; it depends
   on the length to mass ratio.

Answer: (b)
   MC   Easy

11.13: A simple pendulum consists of a 0.25-kg spherical mass attached to a massless string. When the mass is displaced slightly from its equilibrium position and released, the pendulum swings back and forth with a frequency of 2 Hz. What frequency would have resulted if a 0.50-kg mass (same diameter sphere) had been attached to the string instead?
   a) 1 Hz
   c) 1.41 Hz
   b) 2 Hz
   d) None of the above

Answer: (b)
   MC   Easy

11.14: The mass of a mass-and-spring system is displaced 10 cm from its equilibrium position and released. A frequency of 4 Hz is observed. What frequency would be observed if the mass had been displaced only 5 cm and then released?
   a) 2 Hz
   c) 8 Hz
   b) 4 Hz
   d) None of the above

Answer: (b)
   MC   Easy

11.15: If you take a given pendulum to the moon, where the acceleration of gravity is less than on earth, the resonant frequency of the pendulum will
   a) increase.
   b) decrease.
   c) not change.
   d) either increase or decrease; it depends on its length to mass ratio.

Answer: (b)
   MC   Easy

11.16: A mass is attached to a vertical spring and bobs up and down between points A and B. Where is the mass located when its kinetic energy is a maximum?
   a) At either A or B
   b) Midway between A and B
   c) One-fourth of the way between A and B
   d) None of the above

Answer: (b)
   MC   Moderate

11.17: A mass is attached to a vertical spring and bobs up and down between points A and B. Where is the mass located when its kinetic energy is a minimum?
   a) At either A or B
   b) Midway between A and B
   c) One-fourth of the way between A and B
   d) None of the above

Answer: (a)
   MC   Moderate

11.18: A mass vibrates back and forth from the free end of an ideal spring (k = 20 N/m) with an amplitude of 0.25 m. What is the maximum kinetic energy of this vibrating mass?
a) 2.5 J
b) 1.25 J
c) 0.625 J
d) It is impossible to give an answer since kinetic energy cannot be determined without knowing the object's mass.

Answer: (c)
MC   Moderate

11.19: Two masses, A and B, are attached to different springs. Mass A vibrates with an amplitude of 8 cm at a frequency of 10 Hz and mass B vibrates with an amplitude of 5 cm at a frequency of 16 Hz. How does the maximum speed of A compare to the maximum speed of B?
a) Mass A has the greater maximum speed.
b) Mass B has the greater maximum speed.
c) They are equal.

Answer: (c)
MC   Moderate

11.20: A mass vibrates back and forth from the free end of an ideal spring (k = 20 N/m) with an amplitude of 0.30 m. What is the kinetic energy of this vibrating mass when it is 0.30 m from its equilibrium position?
a) Zero
b) 0.90 J
c) 0.45 J
d) It is impossible to give an answer without knowing the object's mass.

Answer: (a)
MC   Easy

11.21: When the length of a simple pendulum is tripled, the time for one complete vibration increases by a factor of $\sqrt{3}$.

Answer: True
TF   Easy

11.22: When the mass of a simple pendulum is tripled, the time required for one complete vibration
a) increases by a factor of 3.
b) does not change.
c) decreases to one-third of its original value.
d) decreases to $1/\sqrt{3}$ of its original value

Answer: (b)
MC   Easy

11.23: A mass is attached to a spring. It oscillates at a frequency of $4/\pi$ Hz when displaced a distance of 2 cm from equilibrium and released. What is the maximum velocity attained by the mass?
a) 0.02 m/s   b) 0.04 m/s
c) 0.08 m/s   d) 0.16 m/s
e) 0.32 m/s

Answer: (d)
MC   Moderate

11.24: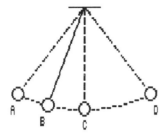
A mass on the end of a massless spring undergoes SHM. Where is the instantaneous acceleration of the mass greatest?
a) A and C   b) B   c) C   d) A and D

Answer: (d)
MC   Moderate

11.25: A simple pendulum consists of a mass M attached to a weightless string of length L. For this system, when undergoing small oscillations
a) the frequency is proportional to the amplitude.
b) the period is proportional to the amplitude.
c) the frequency is independent of the mass M.
d) the frequency is independent of the length L.

Answer: (c)
MC   Moderate

11.26: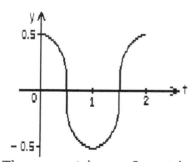
The equation of motion of the wave shown is
a) y = 0.5 sin (πt)   b) y = 0.5 sin (4πt)
c) y = 0.5 cos (πt)   d) y = 0.5 cos (4πt)
e) none of the above.

Answer: (c)
MC   Moderate

152

11.27: A 0.3-kg mass is suspended on a spring. In equilibrium the mass stretches the spring 2 cm downward. The mass is then pulled an additional distance of 1 cm down and released from rest. (a) Calculate the period of oscillation. (b) Calculate the total energy of the system. (c) Write down its equation of motion.

Answer: (a) 0.28 s (b) 0.0074 J (c) y = 0.01 cos(22.1 t)
ES Moderate

11.28: A 2-kg mass is hung from a spring (k = 18 N/m), displaced slightly from its equilibrium position, and released. What is the frequency of its vibration?
a) $1.5/\pi$ Hz
b) $3.0/\pi$ Hz
c) 1.5 Hz
d) None of the above

Answer: (a)
MC Moderate

11.29: Increasing the mass m of a mass-and-spring system causes what kind of change on the resonant frequency of the system?
a) The frequency increases.
b) The frequency decreases.
c) There is no change in the frequency.
d) The frequency increases if the ratio k/m is greater than or equal to 1 and decreases if the ratio k/m is less than 1.

Answer: (b)
MC Easy

11.30: Increasing the spring constant k of a mass-and-spring system causes what kind of change in the resonant frequency of the system? (Assume no change in the system's mass m.)
a) The frequency increases.
b) The frequency decreases.
c) There is no change in the frequency.
d) The frequency increases if the ratio k/m is greater than or equal to 1 and decreases if the ratio k/m is less than 1.

Answer: (a)
MC Easy

11.31: What mass should be attached to a vertical spring (k = 39.5 N/m) so that the natural vibration frequency of the system will be 1.00 Hz?
a) 39.5 kg   b) 6.29 kg   c) 2.00 kg   d) 1.00 kg

Answer: (d)
MC Moderate

11.32: Simple pendulum A swings back and forth at twice the frequency of simple pendulum B. Which statement is correct?
a) Pendulum B is twice as long as A.
b) Pendulum B is twice as massive as A.
c) The length of B is four times the length of A.
d) The mass of B is four times the mass of A.

Answer: (c)
MC   Moderate

11.33: A 2-kg mass is attached to the end of a horizontal spring (k = 50 N/m) and set into simple harmonic motion with an amplitude of 0.1 m. What is the total mechanical energy of this system?
a) 0.02 J   b) 25 J   c) 0.25 J   d) 1.00 J   e) 2.5 J

Answer: (c)
MC   Moderate

11.34: Doubling only the amplitude of a vibrating mass-and-spring system produces what effect on the system's mechanical energy?
a) Increases the energy by a factor of two.
b) Increases the energy by a factor of three.
c) Increases the energy by a factor of four.
d) Produces no change.

Answer: (c)
MC   Easy

11.35: A string of linear density 6 g/m is under a tension of 180 N. What is the velocity of propagation of transverse waves along the string?

Answer: 173 m/s
ES   Moderate

11.36: What is the velocity of propagation if a wave has a frequency of 12 Hz and a wavelength of 3m?
a) 0.25 m/s   b) 4 m/s
c) 36 m/s   d) None of the above

Answer: (c)
MC   Easy

11.37: Find the first three harmonics of a string of linear mass density 2 g/m and length 0.60 m when it is subjected to tension of 50 N.

Answer: 132 Hz, 264 Hz, 395 Hz
ES   Difficult

11.38: A stretched string is observed to have three equal segments in a standing wave driven at a frequency of 480 Hz. What driving frequency will set up a standing wave with four equal segments?

Answer: 640 Hz
ES   Moderate

11.39: Consider a traveling wave on a string of length L, mass M, and tension T. A standing wave is set up. Which of the following is true?
a) The wave velocity depends on M, L, T.
b) The wavelength of the wave is proportional to the frequency.
c) The particle velocity is equal to the wave velocity.
d) The wavelength is proportional to T.

Answer: (a)
MC   Moderate

11.40: A string of linear density 1.5 g/m is under a tension of 20 N. What should its length be if its fundamental resonance frequency is 220 Hz?
a) 0.26 m   b) 0.96 m   c) 1.05 m   d) 1.12 m

Answer: (a)
MC   Difficult

11.41: The velocity of propagation of a transverse wave on a 2-m long string fixed at both ends is 200 m/s. Which one of the following is not a resonant frequency of this string?
a) 25 Hz   b) 50 Hz   c) 100 Hz   d) 200 Hz

Answer: (a)
MC   Moderate

11.42: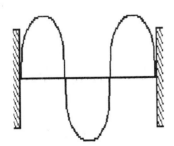
A spring, fixed at both ends, vibrates at a frequency of 12 Hz with a standing transverse wave pattern as shown. What is this spring's fundamental frequency?
a) 36 Hz   b) 16 Hz   c) 8 Hz   d) 4 Hz

Answer: (d)
MC   Moderate

11.43: If one doubles the tension in a violin string, the fundamental
       frequency of that string will increase by a factor of
       a) 2   b) 4
       c) $\sqrt{2}$ d) none of these

Answer: (c)
        MC Moderate

11.44: A mountain climber of mass 60 kg slips and falls 4 m, at which
       time he reaches the end of his elastic safety rope. The rope
       then stretches an additional 2 m before the climber comes to
       rest. Assuming the rope obeys Hooke's Law, what is the spring
       constant of the rope?

Answer: 1760 N/m
        ES Difficult

11.45: A string, fixed at both ends, vibrates at a frequency of 12 Hz
       with a standing transverse wave pattern containing 3 loops. What
       frequency is needed if the standing wave pattern is to contain 4
       loops?
       a) 48 Hz   b) 36 Hz   c) 16 Hz   d) 12 Hz

Answer: (c)
        MC Moderate

11.46: A mass m hanging on a spring has a natural frequency f. If the
       mass is increased to 4m, what is the new natural frequency?
       a) 4f   b) 2f   c) 0.5f   d) 0.25f

Answer: (c)
        MC Easy

11.47: A mass hanging on a spring of constant k has natural frequency f.
       If the spring is changed to one of constant 4k, what is the new
       natural frequency?
       a) 4f   b) 2f   c) 0.5f   d) 0.25f

Answer: (b)
        MC Easy

11.48: A mass on a spring undergoes SHM. When the mass is at its
       maximum displacement from equilibrium, its instantaneous velocity
       a) is maximum.
       b) is less than maximum, but not zero.
       c) is zero.
       d) cannot be determined from the information given.

Answer: (c)
        MC Easy

11.49: A mass on a spring undergoes SHM. When the mass is at maximum displacement from equilibrium, its instantaneous acceleration
a) is a maximum.
b) is less than maximum, but not zero.
c) is zero.
d) cannot be determined from the information given.

Answer: (a)
MC   Easy

11.50: A mass undergoes SHM with an amplitude of 4 cm. The energy is 8 J at this time. The mass is cut in half, and the system is again set in motion with amplitude 4 cm. What is the energy of the system now?
a) 2 J   b) 4 J   c) 8 J   d) 16 J

Answer: (c)
MC   Moderate

11.51: For a periodic process, the number of cycles per unit time is called the
a) amplitude.   b) wavelength.
c) frequency.   d) period.

Answer: (c)
MC   Easy

11.52: The time for one cycle of a periodic process is called the
a) amplitude.   b) wavelength.
c) frequency.   d) period.

Answer: (d)
MC   Easy

11.53: A mass on a spring undergoes SHM. It goes through 8 complete oscillations in 4 s. What is the period?
a) 0.031 s   b) 0.5 s   c) 2 s   d) 32 s

Answer: (b)
MC   Easy

11.54: A mass oscillates on the end of a spring, both on earth and on the moon. Where is the period the greatest?
a) earth
b) the moon
c) same on both earth and moon
d) Cannot be determined from the information given.

Answer: (c)
MC   Moderate

11.55: The number of crests of a wave passing a point per unit time is called the wave's
  a) speed.   b) frequency.
  c) wavelength.   d) amplitude.

Answer: (b)
  MC   Easy

11.56: The distance between successive crests on a wave is called the wave's
  a) speed.   b) frequency.
  c) wavelength.   d) amplitude.

Answer: (c)
  MC   Easy

11.57: In a wave, the maximum displacement of points of the wave from equilibrium is called the wave's
  a) speed.   b) frequency.
  c) wavelength.   d) amplitude.

Answer: (d)
  MC   Easy

11.58: For a wave, the frequency times the wavelength is the wave's
  a) speed.   b) amplitude.   c) intensity.   d) power.

Answer: (a)
  MC   Easy

11.59: The frequency of a wave increases. What happens to the distance between successive crests?
  a) It increases.
  b) It remains the same.
  c) It decreases.
  d) It cannot be determined from the information given.

Answer: (c)
  MC   Easy

11.60: The frequency of a wave increases. What happens to the speed of the wave?
  a) It increases.
  b) It remains the same.
  c) It decreases.
  d) It cannot be determined from the information given.

Answer: (b)
  MC   Easy

11.61: A string of mass m and length L is under tension T. The speed of a wave in the string is v. What will be the speed of a wave in the string if the tension is increased to 2T?

    a) 0.5 T   b) 0.71 T   c) 1.4 T   d) 2 T

Answer: (c)
    MC   Moderate

11.62: A string of mass m and length L is under tension T. The speed of a wave in the string is v. What will be the speed of a wave in the string if the mass of the string is increased to 2m, with no change in length?

    a) 0.5 v   b) 0.71 v   c) 1.4 v   d) 2 v

Answer: (b)
    MC   Easy

11.63: A string of mass m and length L is under tension T. The speed of a wave in the string is v. What will be the speed of a wave in the string if the length is increased to 2L, with no change in mass?

    a) 0.50 v   b) 0.71 v   c) 1.4 v   d) 2 v

Answer: (c)
    MC   Easy

11.64: A wave moves on a string with wavelength λ and frequency f. A second wave on the same string has wavelength 2λ. What is the frequency of the second wave?

    a) 0.5 f
    b) f
    c) 2 f
    d) Cannot be determined from the information given.

Answer: (a)
    MC   Easy

11.65: The most effective way to change the period of a pendulum is to change its

    a) mass.   b) length.   c) weight.   d) amplitude.

Answer: (b)
    MC   Easy

11.66: A 3 kg pendulum is 24.84 m long. What is its period?

    a) 10 s   b) 5 s   c) 3 s   d) 1 s

Answer: (a)
    MC   Easy

11.67: Both pendulum A and B are 3 m long. The period of A is T.
Pendulum A is twice as heavy as pendulum B. What is the period
of B?
   a) 0.71 T   b) T   c) 1.4 T   d) 2 T

Answer: (b)
   MC   Easy

11.68: A pendulum has a period of 2.0 s. What is its length?
   a) 2.0 m   b) 1.0 m   c) 0.70 m   d) 0.50 m

Answer: (b)
   MC   Easy

11.69: What is the frequency of a wave which has a period of 3 ms?
   a) 33.3 Hz   b) 333 Hz   c) 3.33 kHz   d) 33.3 kHz

Answer: (b)
   MC   Easy

11.70: What is the period of a wave with a frequency of 1500 Hz?
   a) 0.67 μs   b) 0.67 ms   c) 0.67 s   d) 6.7 s

Answer: (b)
   MC   Easy

11.71: What is the velocity of a wave that has a wavelength of 4 m and a
frequency of 50 Hz?
   a) 0.005 m/s   b) 0.08 m/s   c) 12.5 m/s   d) 200 m/s

Answer: (d)
   MC   Easy

11.72: What is the wavelength of a 40 kHz wave traveling at 500 m/s?
   a) 0.08 m   b) 12.5 m   c) 80 m   d) 20000 km

Answer: (c)
   MC   Easy

11.73: What is the frequency of a 2.5 m wave traveling at 1400 m/s?
   a) 178 Hz   b) 1.78 kHz   c) 560 Hz   d) 5.6 kHz

Answer: (c)
   MC   Easy

11.1: Match the unit to the physical quantity.
   TA

Sound

12.1: MATCH THE UNIT TO THE PHYSICAL QUANTITY.

12.1: power

Answer: J/s
MA   Easy

12.2: intensity level

Answer: decibel
MA   Easy

12.3: Mach number

Answer: dimensionless
MA   Easy

12.4: intensity

Answer: W/m^2
MA   Easy

12.5: path difference

Answer: meters
MA   Easy

12.6: beat frequency

Answer: Hertz
MA   Easy

12.7: phase difference

Answer: degrees
MA   Easy

12.8: What is the ratio of the speed of sound in air at 0°C to the speed at 100°C?

Answer: 0.85
ES   Moderate

12.9: On a day when the speed of sound in air is 340 m/s, a bat emits a shriek whose echo reaches it 0.025 s later. How far away was the object that reflected back the sound?

Answer: 4.25 m
ES   Easy

12.10: Sound traveling in air at 23°C enters a cold front where the air temperature is 2°C. If the sound frequency is 1500 Hz, determine the wavelength in the warmer air and in the colder air.

Answer: 0.230 m; 0.221 m
ES   Moderate

12.11: If you hear thunder 5 s after seeing a flash of lightning, the distance to the lightning strike is about
    a) 600 m   b) 1200 m   c) 1700 m   d) 2200 m

Answer: (c)
MC   Easy

12.12: When sound passes from air into water
    a) its wavelength does not change.
    b) its frequency does not change.
    c) its velocity does not change.
    d) all of the above are true.

Answer: (b)
MC   Easy

12.13: What happens to the velocity of sound in a gas when the absolute temperature of that gas is doubled?
    a) It doubles.
    b) It quadruples.
    c) It increases to 1.41 times its original value.
    d) None of the above.

Answer: (c)
MC   Moderate

12.14: Sound vibrations with frequency less than 20 Hz are called
    a) infrasonics.   b) ultrasonics.
    c) supersonics.   d) none of the above.

Answer: (a)
MC   Easy

12.15: Sound vibrations with frequencies greater than 20,000 Hz are called
  a) infrasonics.
  b) ultrasonics.
  c) supersonics.
  d) none of the above.

Answer: (b)
  MC   Easy

12.16: As the temperature of the air increases, what happens to the velocity of sound? (Assume that all other factors remain constant.)
  a) It increases.
  b) It decreases.
  c) It does not change.
  d) It increases when atmospheric pressure is high and decreases when the pressure is low.

Answer: (a)
  MC   Easy

12.17: As atmospheric pressure increases, what happens to the velocity of sound? (Assume that all other factors remain constant.)
  a) It increases.
  b) It decreases.
  c) It does not change.
  d) It increases on warm days and decreases on cold days.

Answer: (c)
  MC   Easy

12.18: The wavelength in air of a sound wave of frequency 500 Hz is
  a) 0.68 m   b) 0.75 m   c) 1.47 m   d) 1.80 m   e) 2.00 m

Answer: (a)
  MC   Easy

12.19: Which of the following is a false statement?
  a) Sound waves are longitudinal pressure waves.
  b) Sound can travel through vacuum.
  c) Light travels very much faster than sound.
  d) The transverse waves on a vibrating string are different from sound waves.
  e) "Pitch" (in music) and frequency have approximately the same meaning.

Answer: (b)
  MC   Easy

12.20: If you were to inhale a few breaths from a helium gas balloon, you would probably experience an amusing change in your voice. You would sound like Donald Duck or Alvin the Chipmunk. What is the cause of this curious high-pitched effect?
   a) The helium causes your vocal cords to tighten and vibrate at a higher frequency.
   b) For a given frequency of vibration of your vocal cords, the wavelength of sound is less in helium than it is in air.
   c) Your voice box is resonating at the second harmonic, rather than at the fundamental frequency.
   d) Low frequencies are absorbed in helium gas, leaving the high frequency components, which result in the high, squeaky sound.
   e) Sound travels faster in helium than in air at a given temperature.

Answer: (e)
       MC   Moderate

12.21: What is the intensity of a 70-dB sound?

Answer: 10^{-5} W/m^2
       ES   Easy

12.22: What is the ratio of the intensities of two sounds with intensity levels of 70 dB and 40 dB?

Answer: 1000:1
       ES   Easy

12.23: The intensity at a distance of 6 m from a source that is radiating equally in all directions is 6 x 10^{-10} W/m^2. (a) What is the power level emitted by the source? (b) What is the intensity level in dB?

Answer: (a) 2.7 x 10^{-7} W/m^2 (b) 28 dB
       ES   Moderate

12.24: What is the intensity level of a sound with intensity 10^{-3} W/m^2?
   a) 30 dB   b) 60 dB   c) 90 dB   d) 96 dB

Answer: (c)
       MC   Easy

12.25: You double your distance from a constant sound source that is radiating equally in all directions. What happens to the intensity of the sound? It reduces to
   a) one-half its original value.
   b) one-fourth its original value.
   c) one-sixteenth its original value.
   d) none of the above.

Answer: (b)
       MC   Easy

12.26: You double your distance from a constant sound source that is radiating equally in all directions. What happens to the intensity level of the sound? It drops by
  a) 2 dB.   b) 3 dB.   c) 6 dB.   d) 8 dB.

Answer: (c)
    MC   Moderate

12.27: At a rock concert, if the speakers are placed on the floor instead of higher up on a tower, the intensity of the sound you would hear, from the same distance away, is the same.

Answer: False
    TF   Moderate

12.28: If the phase angle between sound waves from two different point sources is 180°, destructive interference will occur.

Answer: True
    TF   Easy

12.29: The Doppler shift explains
  a) why the siren on a police car changes its pitch as it races past us.
  b) why a sound grows quieter as we move away from the source.
  c) how sonar works.
  d) the phenomenon of beats.
  e) why it is that our hearing is best near 3000 Hz.

Answer: (a)
    MC   Easy

12.30: Suppose that a source of sound is emitting waves uniformly in all directions. If you move to a point twice as far away from the source, the frequency of the sound will be
  a) unchanged.   b) half as great.
  c) one-fourth as great.   d) twice as great.

Answer: (a)
    MC   Easy

12.31: Two adjacent sources each emit a frequency of 800 Hz in air where the velocity of sound is 340 m/s. How much farther back would source 1 have to be moved so an observer in front of the sources would hear no sound?

Answer: 0.213 m
    ES   Moderate

12.32: What is the beat frequency of two sounds that have equal amplitudes and frequencies of 440 Hz and 444 Hz?

Answer: 4 Hz
ES Easy

12.33: The corresponding violin strings on two violins in an orchestra are found to produce a beat frequency of 2 Hz when a frequency of 660 Hz is played. What percentage change in the tension of one of the strings would bring them to the same frequency?

Answer: 0.6%
ES Difficult

12.34: Which of the following increases as a sound becomes louder?
a) Frequency   b) Wavelength
c) Amplitude   d) Period
e) Velocity

Answer: (c)
MC Easy

12.35: Consider the standing wave on a guitar string and the sound wave generated by the guitar as a result of this vibration. What do these two waves have in common?
a) They have the same wavelength.
b) They have the same velocity.
c) They have the same frequency.
d) More than one of the above is true.
e) None of the above is true.

Answer: (c)
MC Moderate

12.36: If a jet plane were to double its MACH-speed, its half-angle will decrease by a factor of:
a) ½   b) 2
c) arc-sin (½)   d) none of the above.

Answer: (d)
MC Difficult

12.37: The half angle of the conical shock wave produced by a supersonic aircraft is 60°. What is the Mach number of the aircraft?
a) 0.87   b) 1.15   c) 1.73   d) 2.0

Answer: (b)
MC Moderate

12.38: A jet flies at a speed of Mach 1.4. What is the half-angle of
the conical shock wave formed?
a) 30°   b) 36°   c) 44°   d) 46°

Answer: (d)
MC   Moderate

12.39: Two pure tones are sounded together and a particular beat
frequency is heard. What happens to the beat frequency if the
frequency of one of the tones is increased?
a) It increases.
b) It decreases.
c) It does not change
d) It could either increase or decrease.

Answer: (d)
MC   Moderate

12.40: An unknown tuning fork is sounded along with a tuning fork whose
frequency is 256 Hz and a beat frequency of 3 Hz is heard. What
is the frequency of the unknown tuning fork?
a) It must be 259 Hz.
b) It must be 253 Hz.
c) It could be either 253 Hz or 259 Hz.
d) It must be 256 Hz.

Answer: (c)
MC   Easy

12.41: In order to produce beats, the two sound waves should have
a) the same amplitude.
b) slightly different amplitudes.
c) the same frequency.
d) slightly different frequencies.

Answer: (d)
MC   Easy

12.42: A sound source (normal frequency of 1000 Hz) approaches a
stationary observer at one-half the speed of sound. The observer
hears a frequency of
a) 2000 Hz.   b) 500 Hz.
c) 1500 Hz.   d) none of the above.

Answer: (a)
MC   Moderate

12.43: An observer approaches a stationary 1000-Hz sound source at twice
the speed of sound. The observer hears a frequency of
a) 4000 Hz.   b) 2000 Hz.
c) 500 Hz.   d) none of the above.

Answer: (d)
MC   Moderate

12.44: An organ pipe open at both ends has a length of 0.80 m. If the velocity of sound in air is 340 m/s, what are the frequencies of the second and third harmonics?

Answer: 425 Hz; 638 Hz
ES   Moderate

12.45: A closed organ pipe of length 0.75 m is played when the speed of sound in air is 340 m/s. What is the fundamental frequency?
    a) 57 Hz   b) 113 Hz   c) 170 Hz   d) 227 Hz

Answer: (b)
MC   Moderate

12.46: Which of the following properties of a sound wave is most closely identified with the "pitch" of a musical note?
    a) Amplitude   b) Wavelength   c) Frequency   d) Phase

Answer: (c)
MC   Easy

12.47: Which of the following is most closely identified with loudness of a musical note?
    a) Frequency   b) Velocity   c) Phase   d) Amplitude

Answer: (d)
MC   Easy

12.48: Only odd harmonics can be produced in an open organ pipe.

Answer: False
TF   Moderate

12.49: The lowest tone to resonate in a closed pipe of length L is 200 Hz. Which of the following frequencies will not resonate in that pipe?
    a) 400 Hz   b) 600 Hz   c) 1000 Hz   d) 1400 Hz

Answer: (a)
MC   Moderate

12.50: The lowest tone to resonate in an open pipe of length L is 400 Hz. What is the frequency of the lowest tone that will resonate in an open pipe of length 2L?
    a) 800 Hz   b) 200 Hz   c) 1600 Hz   d) 100 Hz

Answer: (b)
MC   Moderate

12.51: The lower the frequency of a sound wave, the
    a) lower its velocity.   b) greater its wavelength.
    c) smaller its amplitude.   d) shorter its period.

Answer: (b)
    MC Easy

12.52: The lower the frequency of a sound wave, the
    a) lower its velocity.   b) greater its wavelength.
    c) smaller its amplitude.   d) shorter its period.

Answer: (b)
    MC Easy

12.53: What is the intensity of a 70 dB sound?

Answer: 10^{-5} W/m^2
    ES Easy

12.54: The lower the frequency of a sound wave, the
    a) lower its velocity.   b) greater its wavelength.
    c) smaller its amplitude.   d) shorter its period.

Answer: (b)
    MC Easy

12.55: Compared to the velocity of a 400 Hz sound, the velocity of a 200 Hz sound through air is
    a) twice as great.   b) the same.
    c) one-half as great.   d) none of the above.

Answer: (b)
    MC Easy

12.56: Compared to the wavelength of a 400 Hz sound, the wavelength of a 200 Hz sound in air is
    a) twice as long.   b) the same.
    c) one-half as long.   d) none of the above.

Answer: (a)
    MC Easy

12.57: The wavelengths of the sounds produced by two horns are 6 m and 7 m respectively. What beat frequency is heard when the horns are sounded on a day when the velocity of sound is 340 m/s?

Answer: 8.1 Hz
    ES Moderate

12.58: In general, sound is conducted fastest through
   a) gases.   b) liquids.   c) solids.   d) a vacuum.

Answer: (c)
   MC   Easy

12.59: The lowest tone to resonate in an open pipe of length L is 200
   Hz. Which one of the following frequencies will not resonate in
   the same pipe?
   a) 400 Hz   b) 600 Hz   c) 800 Hz   d) 900 Hz

Answer: (d)
   MC   Moderate

12.60: Consider two pipes of the same length: one is open and the other
   is closed. If the fundamental frequency of the open pipe is 300
   Hz, what is the fundamental frequency of the closed pipe?
   a) 150 Hz   b) 300 Hz   c) 450 Hz   d) 600 Hz

Answer: (a)
   MC   Moderate

12.61: Pressure and displacement waves are
   a) in phase.   b) 45° out of phase.
   c) 90° out of phase.   d) 180° out of phase.

Answer: (c)
   MC   Moderate

12.62: In a resonating pipe which is open at both ends, there
   a) are displacement nodes at each end.
   b) are displacement antinodes at each end.
   c) is a displacement node at one end and a displacement antinode
   at the other end.
   d) none of the above.

Answer: (b)
   MC   Moderate

12.63: In a resonating pipe which is open at one end and closed at the
   other, there
   a) are displacement nodes at each end.
   b) are displacement antinodes at each end.
   c) is a displacement node at the open end and a displacement
   antinode at the closed end.
   d) is a displacement node at the closed end and a displacement
   antinode at the open end.

Answer: (d)
   MC   Moderate

12.64: An open pipe of length L is resonating at its fundamental frequency. Which statement is correct?
a) The wavelength is 2L and there is a displacement node at the pipe's midpoint.
b) The wavelength is 2L and there is a displacement antinode at the pipe's midpoint.
c) The wavelength is L and there is a displacement node at the pipe's midpoint.
d) The wavelength is L and there is a displacement antinode at the pipe's midpoint.

Answer: (a)
MC   Moderate

12.65: Consider an open pipe of length L.   What are the wavelengths of the three lowest tones produced by this pipe?
a) 4L, 2L, L   b) 2L, L, L/2
c) 2L, L, 2L/3   d) 4L, 4L/3, 4L/5

Answer: (c)
MC   Moderate

12.66: Consider a closed pipe of length L.   What are the wavelengths of the three lowest tones produced by this pipe?
a) 4L, 2L, L   b) 2L, L, L/2
c) 2L, L, 2L/3   d) 4L, 4L/3, 4L/5

Answer: (d)
MC   Moderate

12.67: For spherically diverging waves, intensity is proportional to
a) R^2 b) R   c) 1/R   d) $1/R^2$

Answer: (d)
MC   Easy

12.68: A police car has an 800 Hz siren.   It is traveling at 35 m/s on a day when the speed of sound through air is 340 m/s.   The car approaches and passes an observer who is standing along the roadside.   What change of frequency does the observer hear?

Answer: 166 Hz
ES   Moderate

12.69: Your professor strikes a 440 Hz tuning fork and moves away from you and towards the blackboard at a constant velocity of 3 m/s. What beat frequency do you hear if the velocity of sound on that day is 340 m/s?

Answer: 7.76 Hz
ES   Difficult

12.70: Your professor strikes a 440 Hz tuning fork and moves away from you and towards the blackboard at a constant velocity of 3 m/s. What beat frequency does the professor hear if the velocity of sound on that day is 340 m/s?

Answer: 3.92 Hz
ES Difficult

12.71: At a distance of 15 m from a sound source the intensity level is 60 dB. What is the intensity level (in dB) at a point 2 m from the source? Assume that the source radiates equally in all directions.

Answer: 77.5 dB
ES Moderate

12.72: What is the length of the shortest closed pipe that will have a fundamental frequency of 60 Hz on a day when the velocity of sound is 340 m/s?

Answer: 1.42 m
ES Moderate

12.73: You drop a stone into a deep well and hear the splash 2.5 s later. How deep is the well? (Ignore air resistance, and assume the velocity of sound is 340 m/s.)

Answer: 28.6 m
ES Moderate

12.74: On a 30°C day, there is an explosion. The sound is heard 3.4 s after seeing the flash. How far away was the explosion?
   a) 103 m   b) 1125 m   c) 1156 m   d) 1188 m

Answer: (d)
MC Moderate

12.75: Sonar is used to determine the speed of an object. A 40 kHz signal is sent out, and a 42 kHz signal is returned. Assume the speed of sound is 345 m/s. How fast is the object moving?
   a) 6.9 m/s   b) 8.4 m/s   c) 331 m/s   d) 346.8 m/s

Answer: (b)
MC Moderate

12.76: A train is traveling toward you at 120 km/h. The train blows its 400 Hz whistle. Take the speed of sound to be 340 m/s. What frequency do you hear?
   a) 443.5 Hz   b) 439.2 Hz   c) 364.3 Hz   d) 360.8 Hz

Answer: (a)
MC Moderate

12.77: A train is traveling away from you at 120 km/h. It blows its 400 Hz whistle. Take the speed of sound to be 340 m/s. What frequency do you hear?
   a) 443.5 Hz   b) 439.2 Hz   c) 364.3 Hz   d) 360.8 Hz

Answer: (c)
   MC   Moderate

12.78: You are moving at 120 km/h toward a stationary train. The train blows its 400 Hz whistle. Take the speed of sound to be 340 m/s. What frequency do you hear?
   a) 443.5 Hz   b) 439.2 Hz   c) 364.3 Hz   d) 360.8 Hz

Answer: (b)
   MC   Moderate

12.79: You are moving at 120 km/h away from a stationary train. The train blows its 400 Hz whistle. Take the speed of sound to be 340 m/s. What frequency do you hear?
   a) 443.5 Hz   b) 439.2 Hz   c) 364.3 Hz   d) 360.8 Hz

Answer: (d)
   MC   Moderate

12.80: A train is traveling away from you at 120 km/h. It blows its whistle, and you hear a tone of 400 Hz. Take the speed of sound to be 340 m/s. What is the actual frequency of the whistle?
   a) 443.5 Hz   b) 439.2 Hz   c) 364.3 Hz   d) 360.8 Hz

Answer: (b)
   MC   Moderate

12.81: The third harmonic of a complex tone has a frequency of 1200 Hz. What is the frequency of the fourth harmonic?
   a) 400 Hz   b) 900 Hz   c) 1600 Hz   d) 4800 Hz

Answer: (c)
   MC   Easy

12.82: You shout at a cliff, and hear the echo in 4.0 s. The temperature is 0°C. How far away is the cliff?
   a) 662 m   b) 680 m   c) 1324 m   d) 1760 m

Answer: (a)
   MC   Moderate

12.83: You move slowly toward a speaker emitting a pure tone. What characteristic of the sound increases?
   a) frequency   b) amplitude   c) wavelength   d) period

Answer: (b)
   MC   Easy

Temperature and Kinetic Energy

13.1:   **MATCH THE UNIT TO THE PHYSICAL QUANTITY.**

13.1:   temperature of triple point of water

Answer: 273.16°K
         MA   Easy

13.2:   coefficient of volume expansion

Answer: 1/°C
         MA   Easy

13.3:   universal gas constant

Answer: J/mol·K
         MA   Easy

13.4:   boiling point of water

Answer: 373.16°K
         MA   Easy

13.5:   Avogadro's number

Answer: molecules/mole
         MA   Easy

13.6:   9/5 of a Fahrenheit degree

Answer: 1 C°
         MA   Easy

13.7:   Boltzmann's constant

Answer: J/K
         MA   Easy

13.8:   dew point

Answer: °C
         MA   Easy

13.9:   Which temperature scale never gives negative temperatures?
         a) Kelvin   b) Fahrenheit   c) Celsius

Answer: (a)
         MC   Easy

13.10: Which two temperature changes are equivalent?
   a) 1 K° = 1 F°   b) 1 F° = 1 C°   c) 1 C° = 1 K°

Answer: (c)
   MC   Easy

13.11: The temperature in your classroom is approximately
   a) 68°K   b) 68°C   c) 50°C   d) 295°K

Answer: (d)
   MC   Moderate

13.12: Express 45°C in °F.
   a) 25°F   b) 57°F   c) 81°F   d) 113°F

Answer: (d)
   MC   Easy

13.13: At what temperature are the numerical readings on the Fahrenheit and Celsius scales the same?

Answer: -40°
   ES   Moderate

13.14: Express your body temperature (98.6°F) in Celsius degrees.
   a) 37°C   b) 45.5°C   c) 66.6°C   d) 72.6°C

Answer: (a)
   MC   Easy

13.15: A temperature change of 20°C corresponds to a temperature change of
   a) 68°F   b) 11.1°F
   c) 36°F   d) none of the above

Answer: (c)
   MC   Easy

13.16:

A bimetallic strip, consisting of metal G on the top and metal H on the bottom, is rigidly attached to a wall at the left. The coefficient of linear thermal expansion for metal G is greater than that of metal H. If the strip is uniformly heated, it will
   a) curve upward.
   b) curve downward.
   c) remain horizontal, but get longer.
   d) bend in the middle.

Answer: (b)
   MC   Moderate

13.17: A container of an ideal gas at 1 atm is compressed to one-third its volume, with the temperature held constant. What is its final pressure?

Answer: 3 atm
ES Easy

13.18: In order to double the average speed of the molecules in a given sample of gas, the temperature (measured in Kelvins) must
    a) quadruple.
    b) reduce to one-fourth its original value.
    c) reduce to one-half its original value.
    d) triple.
    e) double.

Answer: (a)
MC Moderate

13.19: How many water molecules are there in 36 g of water? Express your answer as a multiple of Avogadro's number N_A. (The molecular structure of a water molecule is H_2O.)
    a) 36 N_A b) 2 N_A
    c) 18 N_A d) None of the above

Answer: (b)
MC Easy

13.20: How many mol are there in 2 kg of copper? (The atomic weight of copper is 63.5 and its specific gravity is 8.9.)

Answer: 31.5 mol
ES Easy

13.21: A constant pressure gas thermometer is initially at 28°C. If the volume of gas increases by 10%, what is the final Celsius temperature?

Answer: 58°C
ES Moderate

13.22: The temperature of an ideal gas increases from 2°C to 4°C while remaining at constant pressure. What happens to the volume of the gas?
    a) It decreases slightly.
    b) It decreases to one-half its original volume.
    c) It more than doubles.
    d) It doubles.
    e) It increases slightly.

Answer: (e)
MC Easy

13.23: Both the pressure and volume of a given sample of an ideal gas double. This means that its temperature in Kelvins must
a) double.
b) quadruple.
c) reduce to one-fourth its original value.
d) remain unchanged.
e) reduce to one-half its original value.

Answer: (b)
MC   Easy

13.24: If the pressure acting on an ideal gas at constant temperature is tripled, its volume is
a) reduced to one-third.
b) increased by a factor of three.
c) increased by a factor of two.
d) reduced to one-half.
e) none of the other choices is correct.

Answer: (a)
MC   Easy

13.25: Two liters of a perfect gas are at $0°C$ and 1 atm.   If the gas is nitrogen, N_2, determine (a) the number of mol (b) the number of molecules (c) the mass of the gas.

Answer: (a)   0.089 mol, (b) 5.34 x 10^{22}, (c) 2.5 g
ES   Moderate

13.26: What is the average separation between air molecules at STP?

Answer: 3.34 x 10^{-7} cm
ES   Difficult

13.27: A sample of a diatomic ideal gas occupies 33.6 L under standard conditions.   How many mol of gas are in the sample?
a) 0.75   b) 3.0
c) 1.5   d) None of the above

Answer: (c)
MC   Moderate

13.28: A container is filled with a mixture of helium and oxygen gases. A thermometer in the container indicates that the temperature is $22°C$.   Which gas molecules have the greater average speed?
a) The helium molecules do because they are monatomic.
b) It is the same for both because the temperatures are the same.
c) The oxygen molecules do because they are more massive.
d) The helium molecules do because they are less massive.
e) The oxygen molecules do because they are diatomic.

Answer: (d)
MC   Moderate

178

13.29: A container is filled with a mixture of helium and oxygen gases. A thermometer in the container indicates that the temperature is 22°C. Which gas molecules have the greater average kinetic energy?
a) It is the same for both because the temperatures are the same.
b) The oxygen molecules do because they are diatomic.
c) The helium molecules do because they are less massive.
d) The helium molecules do because they are monatomic.
e) The oxygen molecules do because they are more massive.

Answer: (a)
MC   Moderate

13.30: Oxygen molecules are 16 times more massive than hydrogen molecules. At a given temperature, the average molecular kinetic energy of oxygen, compared to hydrogen
a) is greater.
b) is less.
c) is the same.
d) cannot be determined since pressure and volume are not given.

Answer: (c)
MC   Moderate

13.31: Oxygen molecules are 16 times more massive than hydrogen molecules. At a given temperature, how do their average molecular speeds compare? The oxygen molecules are moving
a) 4 times faster.   b) at 1/4 the speed.
c) 16 times faster.   d) at 1/16 the speed.

Answer: (b)
MC   Moderate

13.32: A sample of an ideal gas is slowly compressed to one-half its original volume with no change in temperature. What happens to the average speed of the molecules in the sample?
a) It does not change.   b) It doubles.
c) It halves.   d) None of the above.

Answer: (a)
MC   Easy

13.33: A sample of an ideal gas is heated and its Kelvin temperature doubles. What happens to the average speed of the molecules in the sample?
a) It does not change.   b) It doubles.
c) It halves.   d) None of the above.

Answer: (d)
MC   Moderate

13.34: The number of molecules in one mole of a substance
    a) depends on the molecular weight of the substance.
    b) depends on the atomic weight of the substance.
    c) depends on the density of the substance.
    d) is the same for all substances.

Answer: (d)
    MC Easy

13.35: A steel cable is 20 m long when the temperature is 20°C. What will be its length when the temperature drops to 0°C? (The coefficient of thermal expansion of steel is 12×10^{-6} $(°K)^{-1}$).

Answer: 19.9952 m
    ES Moderate

13.36: The coefficient of linear expansion for aluminum is 1.80×10^{-6} $(°K)^{-1}$. What is its coefficient of volume expansion?
    a) 9.0×10^{-6} $(°K)^{-1}$.
    b) 5.8×10^{-18} $(°K)^{-1}$.
    c) 5.4×10^{-6} $(°K)^{-1}$.
    d) 3.6×10^{-6} $(°K)^{-1}$.
    e) None of the above choices is correct.

Answer: (c)
    MC Moderate

13.37: A mercury thermometer has a bulb of volume 0.100 cm^3 at 10°C. The capillary tube above the bulb has a cross-sectional area of 0.012 mm^2. The volume thermal expansion coefficient of mercury is 1.80×10^{-4} $(°K)^{-1}$. How much will the mercury rise when the temperature rises by 30°C?

Answer: 45 mm
    ES Difficult

13.38: When the engine of your car heats up, the spark plug gap will
    a) increase.
    b) decrease.
    c) remain unchanged.
    d) decrease at first and then increase later, so that the two effects cancel once the engine reaches operating temperature.
    e) none of the above is true.

Answer: (a)
    MC Moderate

13.39: Consider a flat steel plate with a hole through its center. When the plate's temperature is increased, the hole will

a) expand only if it takes up more than half the plate's surface area.
b) contract if it takes up less than half the plate's surface area.
c) always contract.
d) always expand.

Answer: (d)
MC   Moderate

13.40: By how much will a slab of concrete 18 m long contract when the temperature drops from 24°C to -16°C? (The coefficient of linear thermal expansion for concrete is 10^{-5} per degree C.)
a) 0.5 cm   b) 0.7 cm   c) 1.2 cm   d) 1.5 cm

Answer: (b)
MC   Moderate

13.41: A bolt hole in a brass plate has a diameter of 1.2 cm at 20°C. What is the diameter of the hole when the plate is heated to 220°C? (The coefficient of linear thermal expansion for brass is 19×10^{-6} per degree C.)
a) 1.205 cm   b) 1.195 cm   c) 1.200 cm   d) 1.210 cm

Answer: (a)
MC   Moderate

13.42: When a lake freezes during the winter, the water at the bottom of the lake is at a higher temperature than the water above it.

Answer: True
TF   Easy

13.43: The surface water temperature on a large, deep lake is 3°C. A sensitive temperature probe is lowered several m into the lake. What temperature will the probe record?
a) A temperature warmer than 3°C
b) A temperature less than 3°C
c) A temperature equal to 3°C

Answer: (a)
MC   Easy

13.44: The volume coefficient of thermal expansion for gasoline is 950 x 10^{-6} per degree C. By how much does the volume of 1 L of gasoline change when the temperature rises from 20°C to 40°C?
a) 6 cm^3 b) 12 cm^3 c) 19 cm^3 d) 37 cm^3

Answer: (c)
MC   Easy

13.45: At what temperature is the average kinetic energy of an atom in helium gas equal to 6.21 x 10^{-21} J?

Answer: 300°K
ES   Easy

13.46: Supersaturation occurs in air when the
a) relative humidity is 100% and the temperature increases.
b) relative humidity is less than 100% and the temperature increases.
c) relative humidity is less 100% and the temperature decreases.
d) relative humidity is 100% and the temperature decreases.

Answer: (d)
MC   Moderate

13.47: Express 72°F in °C.

Answer: 22.2°C
ES   Easy

13.48: Express 30°C in °F.

Answer: 86°F
ES   Easy

13.49: A change of 1°C is larger than a change of 1°F.

Answer: True
TF   Easy

13.50: A temperature of 68°F corresponds to a temperature of
a) 20°C   b) 180.5°C
c) 35.8°C   d) none of the above

Answer: (a)
MC   Easy

13.51: A temperature of -40°C corresponds to a temperature of
a) -54.2°F   b) 4.44°F
c) -40°F   d) None of the above.

Answer: (c)
MC   Easy

13.52: Consider two equal volumes of gas at a given temperature and pressure. One gas, oxygen, has a molecular mass of 32. The other gas, nitrogen, has a molecular mass of 28. What is the ratio of the number of oxygen molecules to the number of nitrogen molecules?

    a) 32:28   b) 28:32

    c) 1:1   d) None of the above.

Answer: (c)

    MC   Moderate

13.53: A given sample of carbon dioxide (CO_2) contains 3.01×10^{23} molecules. What is the mass of this sample?

    a) 44 g   b) 22 g

    c) 22 kg   d) None of the above.

Answer: (b)

    MC   Moderate

13.54: A given sample of carbon dioxide (CO_2) contains 3.01×10^{23} molecules at STP. What volume does this sample occupy?

    a) 11.2 L   b) 22.4 L

    c) 44.8 L   d) None of the above.

Answer: (a)

    MC   Moderate

13.55: The dimension of PV is

    a) [force]   b) [energy]

    c) [energy/temperature]   d) None of the above.

Answer: (b)

    MC   Easy

13.56: If you double both the pressure and absolute temperature of an ideal gas, the volume of the gas will

    a) also double.   b) not change.

    c) be reduced to one-half.   d) quadruple.

Answer: (b)

    MC   Easy

13.57: A sample of helium (He) occupies 44.8 L at STP. What is the mass of the sample?

    a) 2 g   b) 4 g

    c) 8 g   d) None of the above.

Answer: (c)

    MC   Moderate

13.58: A mole of diatomic oxygen molecules and a mole of diatomic nitrogen molecules at STP have
   a) the same average molecular speeds.
   b) the same number of molecules.
   c) the same diffusuion rates.
   d) all of the above.

Answer: (b)
   MC   Moderate

13.59: According to Boyle's Law, PV = constant for a given temperature. As a result, an increase in volume corresponds to a decrease in pressure. This happens because the molecules
   a) collide with each other more frequently.
   b) move slower on the average.
   c) strike the container wall less often.
   d) transfer less energy to the walls of the container each time they strike it.

Answer: (c)
   MC   Moderate

13.60: A container holds N molecules of an ideal gas at a given temperature. If the number of molecules in the container is increased to 2N with no change in temperature or volume, the pressure in the container
   a) doubles.   b) remains constant.
   c) is cut in half.   d) None of the above.

Answer: (b)
   MC   Easy

13.61: The molecular mass of oxygen molecules is 32, and the molecular mass of nitrogen molecules is 28. If these two gases are at the same temperature, the ratio of nitrogen's rms speed to that of oxygen is
   a) $(8)^{1/2}:(7)^{1/2}$ b) 8:7
   c) $(7)^{1/2}:(8)^{1/2}$ d) None of the above.

Answer: (a)
   MC   Moderate

13.62: The molecular mass of nitrogen is 14 times greater than that of hydrogen. If the molecules in these two gases have the same rms speed, the ratio of hydrogen's absolute temperature to that of nitrogen is
   a) $(14)^{1/2}:1$ b) $1:(14)^{1/2}$
   c) 1:14   d) None of the above.

Answer: (c)
   MC   Moderate

13.63: The average molecular kinetic energy of a gas can be determined by knowing only
a) the number of molecules in the gas.
b) the volume of the gas.
c) the pressure of the gas.
d) the temperature of the gas.

Answer: (d)
MC   Easy

13.64: A mixture of gases contains 15 g of H_2, 14 g of N_2, and 44 g of CO_2. The mixture is in a 40 L sealed container which is 20°C. What is the pressure in the container?

Answer: 5.48×10^5 N/m^2
ES   Moderate

13.65: A cylinder contains 16 g of helium gas at STP. How much energy is needed to raise the temperature of this gas to 20°C?

Answer: 978 J
ES   Moderate

13.66: At what temperature would the rms speed of H_2 molecules equal 11,200 m/s (the earth's escape speed)?

Answer: 1.01×10^4 K
ES   Moderate

13.67: A 25 L container holds hydrogen gas at a gauge pressure of 0.25 atm and a temperature of 0°C. What mass of hydrogen is in this container?

Answer: 2.78 g
ES   Moderate

13.68: Convert 14°C to °F.
a) 57.2°F   b) 46°F   c) 39.7°F   d) 60°F

Answer: (a)
MC   Easy

13.69: Convert 14°F to °C.
a) -18°C   b) -10°C   c) 46°C   d) 57.2°C

Answer: (b)
MC   Easy

13.70: Convert 14K to °C.
a) 46°C   b) 287°C   c) 25°C   d) -259°C

Answer: (d)
MC   Easy

13.71: Convert 14°C to K.
a) 46 K   b) 100 K   c) 287 K   d) 474 K

Answer: (c)
MC   Easy

13.72: Convert 14°F to K.
a) 263 K   b) 287 K   c) -10 K   d) 474 K

Answer: (a)
MC   Moderate

13.73: Convert 14 K to °F.
a) 287°F   b) -434°F   c) -259°F   d) 474°F

Answer: (b)
MC   Moderate

13.74: A 100 cm long steel rod cools from 150°C to 40°C. Steel has a
linear expansion coefficient of 12×10^{-6} $(°C)^{-1}$. What is the
change in the rod's length?
a) 132 cm   b) 13.2 cm   c) 1.32 cm   d) 0.132 cm

Answer: (d)
MC   Easy

13.75: An aluminum rod 17.4 cm long at 20°C is heated to 100°C. What is
its new length? Aluminum has a linear expansion coefficient of
25×10^{-6} $(°C)^{-1}$.
a) 17.435 cm   b) 17.365 cm   c) 0.348 cm   d) 0.0348 cm

Answer: (a)
MC   Moderate

13.76: 20.00 cm of space is available. How long a piece of brass at
20°C can be put there and still fit at 200°C? Brass has a linear
expansion coefficient of 19×10^{-6} $(°C)^{-1}$.
a) 19.93 cm   b) 19.69 cm   c) 19.50 cm   d) 19.09 cm

Answer: (a)
MC   Difficult

13.77: A 5 cm diameter steel shaft has 0.1 mm clearance all around its bushing at 20°C. If the bushing temperature remains constant, at what temperature will the shaft begin to bind? Steel has a linear expansion coefficient of 12×10^{-6} $(°C)^{-1}$.
   a) 353°C   b) 333°C   c) 53°C   d) 680°C

Answer: (a)
   MC   Difficult

13.78: 1 L of water at 20°C will occupy what volume at 80°C? Water has a volume expansion coefficient of 210×10^{-6} $(°C)^{-1}$.
   a) 1.6 L   b) 1.013 L   c) 0.987 L   d) 0.9987 L

Answer: (b)
   MC   Easy

13.79: 400 cm^3 of mercury at 0°C will expand to what volume at 50°C? Mercury has a volume expansion coefficient of 180×10^{-6} $(°C)^{-1}$.
   a) 450 cm^3 b) 409.7 cm^3 c) 403.6 cm^3 d) 401.8 cm^3

Answer: (c)
   MC   Easy

13.80: For mecury to expand from 4.0 cm^3 to 4.1 cm^3, what change in temperature is necessary? Mercury has a volume expansion coefficient of 180×10^{-6} $(°C)^{-1}$.
   a) 400°C   b) 139°C   c) 14°C   d) 8.2°C

Answer: (b)
   MC   Moderate

13.81: An ideal gas occupies 4 L at 20°C. What vclume will it occupy at 40°C if the pressure remains constant?
   a) 42.7 cm^3 b) 4.27 L   c) 8.0 L   d) 2.0 L

Answer: (b)
   MC   Easy

13.82: An ideal gas occupies 600 cm^3 at 20°C. At what temperature will it occupy 1200 cm^3 if the pressure remains constant?
   a) 10°C   b) 40°C   c) 100°C   d) 313°C

Answer: (d)
   MC   Easy

13.83: An ideal gas occupies 200 L at an absolute pressure of 300 kPA. Find the absolute pressure if the volume changes to 750 L and the temperature remains constant.
   a) 80 kPa   b) 500 kPa   c) 750 kPa   d) 1125 kPa

Answer: (a)
   MC   Easy

13.84: An ideal gas occupies 400 cm^3 at an absolute pressure of 250 kPa. If the volume is changed to 100 cm^3 at constant temperature, what will be the new gauge pressure?
a) 1000 kPa   b) 899 kPa   c) 164 kPa   d) 62.5 kPa

Answer: (b)
MC Moderate

13.85: A 100 cm^3 container has 4 g of ideal gas in it at 250 kPa. If the volume is changed to 50 cm^3 and the temperature remains constant, what is its new density?
a) 400 kg/m^3 b) 250 kg/m^3 c) 80 kg/m^3 d) 50 kg/m^3

Answer: (c)
MC Moderate

13.86: An ideal gas has a density of 1.75 kg/m^3 at a gauge pressure of 160 kPa. What must be the gauge pressure if a density of 1.0 kg/m^3 is desired at the same temperature?
a) 356 kPa   b) 280 kPa   c) 91 kPa   d) 48 kPa

Answer: (d)
MC Moderate

13.87: 500 cm^3 of ideal gas at 40°C and 200 kPa (absolute) is compressed to 250 cm^3 and cooled to 20°C. What is the final absolute pressure?
a) 748 kPa   b) 374 kPa   c) 200 kPa   d) 100 kPa

Answer: (b)
MC Moderate

13.88: 1500 cm^3 of ideal gas at STP is cooled to -20°C and put into a 1000 cm^3 container. What is the final gauge pressure?
a) 11.1 kPa   b) 39.5 kPa   c) 112.5 kPa   d) 141 kPa

Answer: (b)
MC Moderate

Heat

14.1: **MATCH THE UNIT TO THE PHYSICAL QUANTITY.**

14.1: Stefan-Boltzmann constant

Answer: $W/m^2\text{-}K^4$
MA   Easy

14.2: internal energy

Answer: joules
MA   Easy

14.3: heat of combustion

Answer: kcal/kg
MA   Easy

14.4: specific heat capacity

Answer: kcal/kg-°C
MA   Easy

14.5: emissivity

Answer: dimensionless
MA   Easy

14.6: English unit of heat

Answer: Btu
MA   Easy

14.7: thermal conductivity

Answer: kcal/m-s-°C
MA   Easy

14.8: The units for temperature can be converted into units for heat.

Answer: False
TF   Easy

14.9: A cup of water is scooped up from a swimming pool of water. Compare the temperature T and the internal energy U of the water, in both the cup and the swimming pool.
a) T_{Pool} is greater than T_{Cup}, and the U is the same.
b) T_{Pool} is less than T_{Cup}, and the U is the same.
c) T_{Pool} is equal to T_{Cup}, and U_{Pool} is greater than U_{Cup}.
d) T_{Pool} is equal to T_{Cup}, and U_{Pool} is less than U_{Cup}.

Answer: (c)
MC   Easy

14.10: Which of the following is the smallest unit of heat energy?
a) Calorie   b) Kilocalorie
c) Btu   d) Joule

Answer: (d)
MC   Easy

14.11: Gasoline yields 4.8×10^7 joules per kg when burned. The density of gasoline is approximately the same as that of water, and 1 gal = 3.8 L. How much energy does your car use on a trip of 100 mi if you get 25 mi per gallon?

Answer: 7.3×10^8 J
ES   Moderate

14.12: What is the power output of a hot air furnace that produces heat at the rate of 160,000 BTU/hr?
a) 1.6 kW   b) 22 kW   c) 47 kW   d) 80 kW

Answer: (c)
MC   Easy

14.13: The water flowing over Niagara Falls drops a distance of 50 m. Assuming that all the gravitational energy is converted to thermal energy, by what temperature does the water rise?

Answer: 0.12°C
ES   Moderate

14.14: A person tries to heat up her bath water by adding 5 L of water at 80°C to 60 L of water at 30°C. What is the final temperature of the water?

Answer: 34°C
ES   Moderate

14.15: A 200-L electric water heater uses 2 kW. Assuming no heat loss, how long would it take to heat water in this tank from 23°C to 75°C?

Answer: 21770 s
ES   Moderate

14.16: In grinding a steel knife blade (specific heat = 0.11 cal/g-°C), the metal can get as hot as 400°C. If the blade has a mass of 80 g, what is the minimum amount of water needed at 20°C if the water is not to rise above the boiling point when the hot blade is quenched in it?

Answer: 33 g
ES   Difficult

14.17: It is a well-known fact that water has a higher specific heat capacity than iron. Now, consider equal masses of water and iron that are initially in thermal equilibrium. The same amount of heat, 30 calories, is added to each. Which statement is true?
  a) They remain in thermal equilibrium.
  b) They are no longer in thermal equilibrium; the iron is warmer.
  c) They are no longer in thermal equilibrium; the water is warmer.
  d) It is impossible to say without knowing the exact mass involved and the exact specific heat capacities.

Answer: (b)
MC   Moderate

14.18: Two equal mass objects (which are in thermal contact) make up a system that is thermally isolated from its surroundings. One object has an initial temperature of 100°C and the other has an initial temperature of 0°C. What is the equilibrium temperature of the system, assuming that no phase changes take place for either object? (The hot object has a specific heat capacity that is three times that of the cold object.)
  a) 25°C   b) 50°C
  c) 75°C   d) None of the above

Answer: (c)
MC   Moderate

14.19: A chunk of ice (T = -20°C) is added to a thermally insulated container of cold water (T = 0°C). What happens in the container?
  a) The ice melts until thermal equilibrium is established.
  b) The water cools down until thermal equilibrium is established.
  c) Some of the water freezes and the chunk of ice gets larger.
  d) None of the above.

Answer: (c)
MC   Moderate

14.20: Steam at 100°C will burn your hand equally as severely as water at 100°C will.

Answer: False
TF   Moderate

14.21: Eight grams of water initially at 100°C are poured into a cavity in a very large block of ice initially at 0°C. How many g of ice melt before thermal equilibrium is attained?
   a) 100 g
   b) 10 g
   c) 1 g
   d) An unknown amount; it cannot be calculated without first knowing the mass of the block of ice.

Answer: (b)
   MC   Moderate

14.22: A thermally isolated system is made up of a hot piece of aluminum and a cold piece of copper; the aluminum and the copper are in thermal contact. The specific heat capacity of aluminum is more than double that of copper. Which object experiences the greater temperature change during the time the system takes to reach thermal equilibrium?
   a) The aluminum
   b) The copper
   c) Neither; both experience the same size temperature change.
   d) It is impossible to tell without knowing the masses.

Answer: (d)
   MC   Moderate

14.23: A thermally isolated system is made up of a hot piece of aluminum and a cold piece of copper; the aluminum and the copper are in thermal contact. The specific heat capacity of aluminum is more than double that of copper. Which object experiences the greater magnitude gain or loss of heat during the time the system takes to reach thermal equilibrium?
   a) The aluminum
   b) The copper
   c) Neither; both experience the same size gain or loss of heat.
   d) It is impossible to tell without knowing the masses.

Answer: (c)
   MC   Moderate

14.24: Phase changes occur
   a) as the temperature decreases.
   b) as the temperature increases.
   c) as the temperature remains the same.
   d) all of the above are possible.

Answer: (c)
   MC   Easy

14.25: A block of ice at 0°C is added to a 150-g aluminum calorimeter cup that holds 200 g of water at 10°C. If all but 2 g of ice melt, what was the original mass of the block of ice?
   a) 31.1   b) 38.8   c) 42.0   d) 47.6

Answer: (a)
   MC Difficult

14.26: The heat of fusion of lead is 5.9 kcal/kg, and the heat of vaporization is 207 kcal/kg, and its melting point is 328°C. How much heat is required to melt 50 g of lead initially at 23°C? (The specific heat of lead is 0.031 kcal/kg-°C.)

Answer: 768 cal
   ES Difficult

14.27: The heat of fusion of ice is 80 kcal/kg. When 50 g of ice at 0°C is added to 50 g of water at 25°C, what is the final temperature?

Answer: 0°C
   ES Moderate

14.28: Turning up the flame under a pan of boiling water causes
   a) the water to boil away faster.
   b) the temperature of the boiling water to increase.
   c) both the water to boil away faster and the temperature of the boiling water to increase.
   d) none of the above.

Answer: (a)
   MC Easy

14.29: If heat is added to a pure substance at a steady rate,
   a) its temperature will begin to rise.
   b) it will eventually melt.
   c) it will eventually boil.
   d) more than one of the above is true.
   e) none of the above is true.

Answer: (d)
   MC Easy

14.30: At what rate is the human body radiating energy when it is at 33°C? Take the body surface area to be 1.4 m², and approximate the body as a blackbody.

Answer: 696 W
   ES Easy

14.31: In an electric furnace used for refining steel, the temperature is monitored by measuring the radiant power emitted through a small hole of area 0.5 cm^2. The furnace acts like a blackbody radiator. If it is to be maintained at a temperature of 1650°C, at what level should the power radiated through the hole be maintained?

Answer: 38.8 W
ES Easy

14.32: The effect of using a large fan in a closed room will be to lower the air temperature.

Answer: False
TF Moderate

14.33: The thermal conductivity of concrete is 0.8 W/m-°C and the thermal conductivity of wood is 0.1 W/m-°C. How thick would a solid concrete wall have to be in order to have the same rate of flow through it as an 8-cm thick wall made of solid wood? (Assume both walls have the same surface area.)

Answer: 64 cm
ES Easy

14.34: If you double the absolute temperature of an object, it will radiate energy
    a) 16 times faster.   b) 8 times faster.
    c) 4 times faster.   d) none of the above.

Answer: (a)
MC Easy

14.35: Convection can occur
    a) only in solids.
    b) only in liquids.
    c) only in gases.
    d) only in liquids and gases.
    e) in solids, liquids, and gases.

Answer: (d)
MC Easy

14.36: Consider two neighboring rectangular houses built from the same materials. One of the houses has twice the length, width, and height of the other. Under identical climatic conditions, what would be true about the rate that heat would have to be supplied to maintain the same inside temperature on a cold day? Compared to the small house, the larger house would need heat supplied at
    a) twice the rate.   b) 4 times the rate.
    c) 16 times the rate.   d) none of the above.

Answer: (b)
MC Moderate

14.37: By what primary heat transfer mechanism does the sun warm the earth?
a) Convection
b) Conduction
c) Radiation
d) All of the above in combination

Answer: (c)
MC   Easy

14.38: By what primary heat transfer mechanism does one end of an iron bar become hot when the other end is placed in a flame?
a) Natural convection   b) Conduction
c) Radiation   d) Forced convection

Answer: (b)
MC   Easy

14.39: What temperature exists inside a solar collector (effective collection area of 15 m^2) on a bright sunny day when the outside temperature is +20°C? Assume that the collector is thermally insulated, that the sun irradiates the collector with a power per unit area of 600 W/m^2, and that the collector acts as a perfect blackbody.
a) 73°C   b) 93°C   c) 107°C   d) 154°C

Answer: (b)
MC   Difficult

14.40: The thermal conductivity of aluminum is twice that of brass. Two rods (one aluminum and the other brass) are joined together end to end in excellent thermal contact. The rods are of equal lengths and radii. The free end of the brass rod is maintainedat 0°C and the aluminum's free end is heated to 200°C. If no heat escapes from the sides of the rods, what is the temperature at the interface between the two metals?
a) 76°C   b) 133°C   c) 148°C   d) 155°C

Answer: (b)
MC   Moderate

14.41: A layer of insulating material with thermal conductivity K is placed on a layer of another material of thermal conductivity 2K. The layers have equal thickness. What is the effective thermal conductivity of the composite sheet?
a) 3K   b) 1.5K   c) K/3   d) 2K/3

Answer: (d)
MC   Difficult

14.42: A spaceship is drifting in an environment where the acceleration of gravity is essentially zero. As the air on one side of the cabin is heated by an electric heater, what is true about the convection currents caused by this heating?
a) The hot air around the heater rises and the cooler air moves in to take its place.
b) The hot air around the heater drops and the cooler air moves in to take its place.
c) The convection currents move about the cabin in a random fashion.
d) There are no convection currents.

Answer: (c)
MC   Moderate

14.43: If you double the thickness of a wall built from a homogeneous material, the rate of heat loss for a given temperature difference across the thickness will
a) become one-half its original value.
b) also double.
c) become one-fourth its original value.
d) none of the above.

Answer: (a)
MC   Easy

14.44: When a vapor condenses
a) the temperature of the substance increases.
b) the temperature of the substance decreases.
c) heat energy leaves the substance.
d) heat energy enters the substance.

Answer: (c)
MC   Easy

14.45: In a liquid at a given temperature, the molecules are moving in every direction, some fast, some slowly. Electrical forces of adhesion tend to hold them together. However, occasionally one molecule gains enough energy (as a result of collisions) so that it pulls loose from its neighbors and escapes from the liquid. Which of the following can best be understood in terms of this phenomena?
a) A hot water bottle will do a better job of keeping you warm than will a rock of the same mass heated to the same temperature.
b) When a large steel suspension bridge is built, gaps are left between the girders.
c) If snow begins to fall when you are skiing, you will feel colder than you did before it started to snow.
d) When you step out of a swimming pool and stand in the wind, you will get colder than you would if you stayed out of the wind.
e) Increasing the atmospheric pressure over a liquid will cause the boiling temperature to decrease.

Answer: (d)
MC Moderate

14.46: Which of the following best explains why sweating is important to humans in maintaining suitable body temperature?
a) Moisture on the skin increases thermal conductivity, thereby allowing heat to flow out of the body more effectively.
b) Evaporation of moisture from the skin extracts heat from the body.
c) The high specific heat of water on the skin absorbs heat from the body.
d) Functioning of the sweat glands absorbs energy that otherwise would go into heating the body.
e) None of the above explains the principle on which sweating depends.

Answer: (b)
MC Easy

14.47: In a cold climate, its generally better to wear several layers of clothing than one piece of clothing of the same thickness because air has a smaller thermal conductivity than clothing does.

Answer: True
TF Moderate

14.48: Heat is added to an ideal gas at 20°C. If the internal energy of the gas increases by a factor of three during this process, what is the final temperature?

Answer: 606°C
ES Moderate

14.49: An ideal gas with internal energy U at 200°C is heated to 400°C. Its internal energy then will be
a) still U.   b) 2 U.   c) 1.4 U.   d) 1.2 U.

Answer: (c)
MC Moderate

14.50: An ideal gas at STP is first compressed until its volume is half the initial volume, and then it is allowed to expand until its pressure is half the initial pressure. All of this is done while holding the temperature constant. If the initial internal energy of the gas is U, the final internal energy of the gas will be
a) U.   b) U/3.   c) U/2.   d) 2U.

Answer: (a)
MC Moderate

14.51: The internal energy of an ideal gas depends on
a) its volume.   b) its pressure.
c) its temperature.   d) all of the above.

Answer: (c)
MC Easy

14.52: On his honeymoon, James Joule attempted to explore the relationships between various forms of energy by measuring the rise of temperature of water which had fallen down a waterfall on Mount Blanc. What maximum temperature rise would one expect for a waterfall with a vertical drop of 20 m?

Answer: 0.047°C
ES Moderate

14.53: "Degrees Celsius" can be converted to "joules" or "calories" by the appropriate conversion factor.

Answer: False
TF Easy

14.54: Internal energy is that energy given by the relation:
$$H = E_{total} - KE - PE$$

Answer: False
TF Easy

14.55: Cold is the energy which flows from a cold object to a hot object by virtue of their temperature difference.

Answer: False
TF Easy

14.56: Absolute temperature is proportional to the average kinetic energy of the atoms or molecules making up a system.

Answer: True
    TF   Easy

14.57: Temperature and heat have the same meaning.

Answer: False
    TF   Easy

14.58: An aluminum kettle (mass 1000 g, c = 0.22 kcal/kg °C) holds 400 g of pure water at 20°C. The kettle is placed on a 1000 W electric burner and heated to boiling. Assume that all the heat from the burner heats the kettle and its contents, and that a negligible amount of water evaporates before boiling begins. Calculate the amount of time required to bring the water to boil.
    a) 3.5 min   b) 4.0 min   c) 7.3 min   d) 8.1 min

Answer: (a)
    MC   Difficult

14.59: 1700 J of work is equivalent to how much heat?
    a) 7,116,000 cal   b) 7.116 kcal
    c) 406 cal   d) 406 kcal

Answer: (c)
    MC   Easy

14.60: 14.5 kcal of heat is equivalent to how much work?
    a) 3.46 J   b) 3460 J   c) 60.7 J   d) 60700 J

Answer: (d)
    MC   Easy

14.61: How much heat will flow in 1.0 hour through a 2 m x 2 m section of concrete (k = 2.0 x 10^{-4} kcal/s·m·°C) 10 cm thick if the inside temperature is 21°C and the outside temperature is 4°C?
    a) 0.136 cal   b) 136 cal   c) 490 cal   d) 490 kcal

Answer: (d)
    MC   Moderate

14.62: How long will it take to transfer 1,000,000 cal of heat through a 2 m^2 pane of 0.3 cm thick glass (k = 2.0 x 10^{-4} kcal/s·m·°C) if the temperature differential is 10°C?
    a) 208 hr   b) 20.8 hr   c) 12.5 min   d) 75 s

Answer: (c)
    MC   Moderate

14.63: What is the outside temperature if 4000 kcal of heat is lost through a 4 m^2 pane of 0.3 cm thick glass (k = 2.0 x 10^{-4} kcal/s·m·°C) in 1 hour from a house kept at 20°C?
   a) 0°C   b) 4°C   c) 16°C   d) 24°C

Answer: (c)
   MC   Difficult

14.64: How much heat is needed to raise the temperature of 100 g of lead (c = 0.11 kcal/kc·°C) by 15°C?
   a) 16.5 cal   b) 165 cal   c) 1500 cal   d) 15 kcal

Answer: (b)
   MC   Easy

14.65: 150 kcal of heat raises the temperature of 2 kg of material by 400°F. What is the material's specific heat capacity?
   a) 1.35 kcal/kg·°C   b) 0.75 kcal/kg·°C
   c) 0.38 kcal/kg·°C   d) 0.19 kcal/kg·°C

Answer: (c)
   MC   Moderate

14.66: If 40 kcal of heat is added to 2 kg of water, what is the resulting temperature change?
   a) 80°C   b) 40°C   c) 20°C   d) 0.05°C

Answer: (c)
   MC   Easy

14.67: If 50 g of material at 100°C is mixed with 100 g of water at 0°C, the final temperature is 40°C. What is the specific heat of the material?
   a) 0.33 kcal/kg·°C   b) 0.75 kcal/kg·°C
   c) 1.33 kcal/kg·°C   d) 7.5 kcal/kg·°C

Answer: (c)
   MC   Moderate

14.68: 50 g of lead (c = 0.11 kcal/kg·°C) at 100°C is put into 75 g of water at 0°C. What is the final temperature of the mixture?
   a) 2°C   b) 6.8°C   c) 25°C   d) 50°C

Answer: (b)
   MC   Moderate

14.69: 50 g of copper (c = 0.093 kcal/kg·°C) at 75°C is put into 100 g of water. The final temperature of the mixture is 20°C. What was the initial temperature of the water?
   a) 0°C   b) 4°C   c) 15°C   d) 17.4°C

Answer: (d)
   MC   Moderate

14.70: Ice has a latent heat of fusion of 80 kcal/kg. How much work is required to melt 200 g of ice?
   a) 400 J   b) 160 J   c) 67 J   d) 16 J

Answer: (c)
   MC   Moderate

14.71: How much heat must be removed from steam to change it to liquid?
   a) 540 cal/g   b) 600 cal/g   c) 1 kcal/g   d) 1.8 kcal/g

Answer: (a)
   MC   Easy

14.72: If 2 kg of water at 0°C is to be vaporized, how much heat must be added?
   a) 1080 cal   b) 1080 kcal   c) 1280 cal   d) 1280 kcal

Answer: (d)
   MC   Moderate

14.73: How much heat is required to change 1 g of 0°C ice to 120°C steam?
   a) 48.7 cal   b) 120 cal   c) 730 cal   d) 1505 cal

Answer: (c)
   MC   Moderate

14.74: How much heat needs to be removed from 100 g of 85°C water to make -5°C ice?
   a) 255 cal   b) 8.5 kcal   c) 16.5 kcal   d) 16.8 kcal

Answer: (d)
   MC   Moderate

14.75: How much heat is required to change 100 g of -10°C ice to 150°C steam?
   a) 74.9 kcal   b) 54 kcal   c) 749 cal   d) 594 cal

Answer: (a)
   MC   Difficult

15. The Laws of Thermodynamics

15.1: **MATCH THE FORMULA TO THE PHYSICAL QUANTITY.**

15.1: entropy

Answer: Q/T
MA Easy

15.2: thermal efficiency (non-Carnot)

Answer: $1 - Q_L/Q_H$
MA Easy

15.3: coefficient of performance

Answer: Q_L/W
MA Easy

15.4: Carnot efficiency

Answer: $1 - T_L/T_H$
MA Easy

15.5: change in internal energy

Answer: $Q - W$
MA Easy

15.6: isobaric work

Answer: $p \Delta V$
MA Easy

15.7: An example of a reversible process is driving a car.

Answer: False
TF Easy

15.8: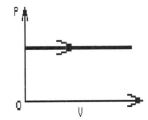

The process shown on the p-V graph is an
a) adiabatic expansion.   b) isothermal expansion.
c) isometric expansion.   d) isobaric expansion.

Answer: (d)
MC   Easy

15.9: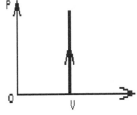

The process shown on the p-V graph is
a) adiabatic.   b) isothermal.
c) isochoric.   d) isobaric.

Answer: (c)
MC   Easy

15.10: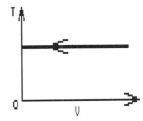

The process shown on the T-V graph is an
a) adiabatic compression.   b) isothermal compression.
c) isochoric compression.   d) isobaric compression.

Answer: (b)
MC   Easy

15.11: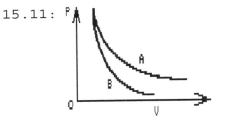

(Two processes are shown on the p-V graph;
one is an adiabat and the other is an
isotherm.)
The process represented by the upper curve
is an isotherm.

Answer: True
TF   Moderate

203

15.12: A gas is allowed to expand at constant pressure as heat is added to it. This process is
    a) isothermal.   b) isochoric.
    c) isobaric.   d) adiabatic.

Answer: (c)
    MC   Easy

15.13: A gas is confined to a rigid container that cannot expand as heat energy is added to it. This process is
    a) isothermal.   b) isochoric.
    c) isobaric.   d) adiabatic.

Answer: (b)
    MC   Easy

15.14: A gas is expanded to twice its original volume with no change in its temperature. This process is
    a) isothermal.   b) isochoric.
    c) isobaric.   d) adiabatic.

Answer: (a)
    MC   Easy

15.15: A gas is quickly compressed in an isolated environment. During the event, the gas exchanged no heat with its surroundings. This process is
    a) isothermal.   b) isochoric.
    c) isobaric.   d) adiabatic.

Answer: (d)
    MC   Easy

15.16: 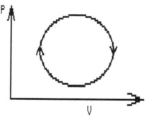 An ideal gas is subjected to one complete cycle of the reversible process shown on the p-V graph. The net work done during this cycle is
    a) positive.   b) negative.   c) zero.

Answer: (a)
    MC   Easy

15.17: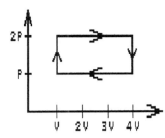

A gas is taken through the cycle illustrated here. During one cycle, how much work is done by an engine operating on this cycle?
 a) pV   b) 2pV   c) 3pV   d) 4pV

Answer: (c)
 MC   Easy

15.18: 200 J of work is done in compressing a gas adiabatically. What is the change in internal energy of the gas?

Answer: 200 J
 ES   Easy

15.19: An ideal gas is compressed isothermally from 30 L to 20 L. During this process, 6 J of energy is expended by the external mechanism that compressed the gas. What is the change of internal energy for this gas?
 a) +6 J   b) zero
 c) -6 J   d) None of the above

Answer: (b)
 MC   Moderate

15.20: When the first law of thermodynamics, $Q = \Delta U + W$, is applied to an ideal gas that is taken through an isothermal process,
 a) $\Delta U = 0$ b) $W = 0$
 c) $Q = 0$ d) none of the above is true

Answer: (a)
 MC   Moderate

15.21: When the first law of thermodynamics, $Q = \Delta U + W$, is applied to an ideal gas that is taken through an isobaric process,
 a) $\Delta U = 0$ b) $W = 0$
 c) $Q = 0$ d) none of the above is true.

Answer: (d)
 MC   Moderate

15.22: When the first law of thermodynamics, $Q = \Delta U + W$, is applied to an ideal gas that is taken through an adiabatic process,
 a) $\Delta U = 0$ b) $W = 0$
 c) $Q = 0$ d) none of the above is true.

Answer: (c)
 MC   Easy

15.23: An ideal gas is compressed to one-half its original volume during
an isothermal process. The final pressure of the gas
a) increases to twice its original pressure.
b) increases to less than twice its original pressure.
c) increases to more than twice its original pressure.
d) does not change.

Answer: (a)
MC   Easy

15.24: A monatomic ideal gas is compressed to one-half its original
volume during an adiabatic process. The final pressure of the
gas
a) increases to twice its original value.
b) increases to less than twice its original value.
c) increases to more than twice its original value.
d) does not change.

Answer: (c)
MC   Moderate

15.25: Consider two cylinders of gas identical in all respects except
that one contains O_2 and the other He. Both hold the same volume
of gas at STP and are closed by a movable piston at one end.
Both gases are now compressed adiabatically to one-third their
original volume. Which gas will show the greater temperature
increase?
a) The O_2
b) The He
c) Neither; both will show the same increase.
d) It's impossible to tell from the information given.

Answer: (b)
MC   Difficult

15.26: Consider two cylinders of gas identical in all respects except
that one contains O_2 and the other He. Both hold the same volume
of gas at STP and are closed by a movable piston at one end.
Both gases are now compressed adiabatically to one-third their
original volume. Which gas will show the greater pressure
increase?
a) The O_2
b) The He
c) Neither; both will show the same increase.
d) Its impossible to tell form the information given.

Answer: (b)
MC   Difficult

15.27: A container of ideal gas at STP undergoes an isothermal expansion
and its entropy changes by 3.66 J/°K. How much work does it do?

Answer: 999 J
ES   Moderate

15.28: What is the change in entropy when 50 g of ice melt at 0°C?

Answer: 4 kcal/°K
ES   Easy

15.29: A piece of metal at 80°C is placed in 1.2 L of water at 72°C. The system is thermally isolated and reaches a final temperature of 75°C. Estimate the approximate change in entropy for this process.

Answer: 2.53 cal/°K
ES   Difficult

15.30: When water freezes, the entropy of the water
a) increases.
b) decreases.
c) does not change.
d) could either increase or decrease; it depends on other factors.

Answer: (b)
MC   Easy

15.31: Is it possible to transfer heat from a hot reservoir to a cold reservoir?
a) No.
b) Yes; this will happen naturally.
c) Yes, but work will have to be done.
d) Theoretically yes, but it hasn't been accomplished yet.

Answer: (b)
MC   Easy

15.32: When a gas expands adiabatically, the work done by the gas is zero.

Answer: False
TF   Easy

15.33: When a gas is compressed isothermally, the work done by the gas is negative.

Answer: True
TF   Moderate

15.34: When a gas is heated isochorically, the work done by the gas is positive.

Answer: False
TF   Moderate

15.35: An ideal gas is expanded isothermally from 20 L to 30 L. During this process, 6 J of energy is expended by the external mechanism that expanded the gas. Which of the following statements is correct?
a) 6 J of energy flow from surroundings into the gas.
b) 6 J of energy flow from the gas into the surroundings.
c) No energy flows into or from the gas since this process is isothermal.
d) None of the above statements is correct.

Answer: (a)
MC Moderate

15.36: A certain amount of a monatomic gas is maintained at constant volume as it is cooled by 50°K. This feat is accomplished by removing 400 J of energy from the gas. How much work is done by the gas?
a) Zero   b) 400 J
c) -400 J   d) None of the above

Answer: (a)
MC Moderate

15.37: A monatomic gas is cooled by 50°K at constant volume when 831 J of energy is removed from it. How many moles of gas are in the sample?
a) 2.50 mol   b) 1.50 mol
c) 1.33 mol   d) None of the above

Answer: (c)
MC Moderate

15.38: The second law of thermodynamics leads us to conclude that
a) the total energy of the universe is constant.
b) disorder in the universe is increasing with the passage of time.
c) it is theoretically possible to convert heat into work with 100% efficiency.
d) the average temperature of the universe is increasing with the passage of time.

Answer: (b)
MC Easy

15.39: A heat engine absorbs 64 kcal of heat each cycle and exhausts 42 kcal. (a) Calculate the efficiency each cycle. (b) Calculate the work done each cycle.

Answer: (a) 34%   (b) 22 kcal
ES Moderate

15.40: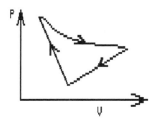

A cyclic process is carried out on an ideal gas such that it returns to its initial state at the end of a cycle. If the process was carried out in a clockwise sense around the enclosed area, as shown on the p-V diagram, then that area represents
a) the internal energy change of the ideal gas.
b) the heat that flows from the ideal gas.
c) the work done by the ideal gas.
d) the work done on the ideal gas.

Answer: (c)
MC   Moderate

15.41: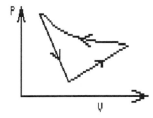

A cyclic process is carried out on an ideal gas such that it returns to its initial state at the end of a cycle. If the process was carried out in a counter-clockwise sense around the enclosed area, as shown on the p-V diagram, then that area represents
a) the heat added to the ideal gas.
b) the internal energy change of the ideal gas.
c) the work done by the ideal gas.
d) the work done on the ideal gas.

Answer: (d)
MC   Moderate

15.42: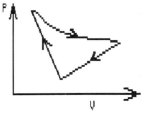

A cyclic process is carried out on an ideal gas such that it returns to its initial state at the end of a cycle. If the process was carried out in a clockwise sense around the enclosed area, as shown on the p-V diagram, then the change of internal energy over the full cycle
a) is positive
b) is negative.
c) is zero.
d) cannot be determined from the information given.

Answer: (c)
MC   Moderate

15.43: A heat engine operating between 40°C and 380°C has an efficiency 60% of that of a Carnot engine operating between the same temperatures. If the engine absorbs heat at a rate of 60 kW, at what rate does it exhaust heat?

Answer: 41.3 kW
ES Difficult

15.44: What is the maximum theoretical efficiency possible for an engine operating between 100°C and 400°C?

A) 25%   (B) 35%   (C) 45%   D) 40%

Answer: 45%
ES Easy

15.45: A Carnot cycle consists of
   a) two adiabats and two isobars.
   b) two isobars and two isotherms.
   c) two isotherms and two isomets.
   d) two adiabats and two isotherms.

Answer: (d)
MC Moderate

15.46: If the theoretical efficiency of a Carnot engine is to be 100%, the heat sink must be
   a) at absolute zero.   b) at 0°C.
   c) at 100°C.   d) infinitely hot.

Answer: (a)
MC Moderate

15.47: What is the theoretical efficiency of a Carnot engine that operates between 600°K and 300°K?
   a) 100%   b) 50%
   c) 25%   d) None of the above

Answer: (b)
MC Easy

15.48: A coal-fired plant generates 600 MW of electric power. The plant uses 4.8×10^6 kg of coal each day. The heat of combustion of coal is 3.3×10^7 J/kg. The steam that drives the turbines is at a temperature of 300°C, and the exhaust water is at 37°C. (a) What is the overall efficiency of the plant for generating electric power? (b) What is the Carnot efficiency? (c) How much thermal energy is wasted each day?

Answer: (a) 33%   (b) 46%   (c) 1.1×10^{14} J
ES Moderate

15.49: One of the most efficient engines built so far has the following
characteristics:
Combustion chamber temperature = 1900°C
Exhaust temperature = 430°C
7×10^9 cal of fuel produces 1.4×10^{10} J of work in one hour.
(a) What is the actual efficiency of this engine? (b) What is
the Carnot efficiency of this engine? (c) What is the power
output, in hp, of this engine?

Answer: (a) 48%   (b) 68%   (c) 5213 hp
ES   Moderate

15.50: Ten joules of heat energy are transferred to a sample of ideal
gas at constant pressure. As a result, the internal energy of
the gas
a) increases by 10 J.
b) increases by less than 10 J.
c) increases by more than 10 J.
d) remains unchanged.

Answer: (b)
MC   Moderate

15.51: Ten joules of heat energy are transferred to a sample of ideal
gas at constant volume. As a result, the internal energy of the
gas
a) increases by 10 J.
b) increases by less than 10 J.
c) increases by more than 10 J.
d) remains unchanged.

Answer: (a)
MC   Moderate

16. Electric Charge and Electric Field

16.1: **MATCH THE UNIT TO THE PHYSICAL QUANTITY.**

16.1: Coulomb constant

Answer: $N-m^2/C^2$
MA Easy

16.2: permittivity

Answer: $C^2/N-m^2$
MA Easy

16.3: charge

Answer: coulomb
MA Easy

16.4: electric field

Answer: N/C
MA Easy

16.5: Electrons carry a
   a) positive charge.   b) negative charge.
   c) neutral charge.   d) variable charge.

Answer: (b)
MC Easy

16.6: Charge is
   a) quantized.   b) conserved.
   c) invariant.   d) all of the above.

Answer: (d)
MC Easy

16.7: Quarks have greater charges than electrons.

Answer: False
TF Moderate

16.8: The ratio of the neutron mass to the electron mass is
   approximately
      a) 2:1   b) 20:1   c) 200:1   d) 2000:1

Answer: (d)
MC Easy

212

16.9: A neutral atom always has
   a) more neutrons than protons.
   b) more protons than electrons.
   c) the same number of neutrons as protons.
   d) the same number of protons as electrons.

Answer: (d)
   MC   Easy

16.10: A glass rod is rubbed with a piece of silk. During the process
   the glass rod acquires a positive charge and the silk
   a) acquires a positive charge also.
   b) acquires a negative charge.
   c) remains neutral.
   d) could either be positively charged or negatively charged. It
   depends on how hard the rod was rubbed.

Answer: (b)
   MC   Easy

16.11: Consider an equilateral triangle of side 20 cm. A charge of $+2\mu C$
   is placed at one vertex and charges of $-4\mu C$ are placed at the
   other two vertices. Determine the magnitude and direction of the
   electric field at the center of the triangle.

Answer: 2.7×10^6 N/C away from the $+2 \mu C$ charge
   ES   Difficult

16.12: A metal sphere of radius 2 cm carries a charge of $3\mu C$. What is
   the electric field 6 cm from the center of the sphere?

Answer: 7.5×10^6 N/C
   ES   Moderate

16.13: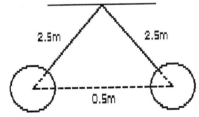

Two volleyballs, each of mass 0.3 kg, are charged by an electrostatic generator. Each is attached to an identical string and suspended from the same point. They repel each other and hang with separation 0.5 m. The length of the string from the point of support to the center of a ball is 2.5 m. Determine the charge on each ball.

Answer: $2.86\mu C$
   ES   Difficult

16.14: Charge $+2q$ is placed at the origin and charge $-q$ is placed at x =
   2a. Where can a third positive charge $+q$ be placed so that the
   force on it is zero?

Answer: x = 6.83 a
   ES   Difficult

16.15: An electron and a proton are separated by a distance of 1 m. What happens to the size of the force on the proton if the electron is moved 0.5 m closer to the proton?
   a) It increases to 4 times its original value.
   b) It increases to 2 times its original value.
   c) It decreases to one-half its original value.
   d) It decreases to one-fourth its original value.

Answer: (a)
   MC   Easy

16.16: The charge carried by one electron is $e = -1.6 \times 10^{-19}$ C. The number of excess electrons necessary to produce a charge of -1.0 C is
   a) 6.25×10^{18} b) 6.25×10^{9}
   c) 1.6×10^{19} d) none of the above.

Answer: (a)
   MC   Easy

16.17: Sphere A carries a net charge and sphere B is neutral. They are placed near each other on an insulated table. Which statement best describes the electrostatic force between them?
   a) There is no force between them since one is neutral.
   b) There is a force of repulsion between them.
   c) There is a force of attraction between them.
   d) The force is attractive if A is charged positively and repulsive if A is charged negatively.

Answer: (c)
   MC   Moderate

16.18:   A point charge of +Q is placed at the center of a square, and a second point charge of -Q is placed at the upper-left corner. It is observed that an electrostatic force of 2 N acts on the positive charge at the center. What is the magnitude of the force that acts on the center charge if a third charge of -Q is placed at the lower-left corner?
   a) Zero   b)
                                       $2\sqrt{2}$ N
   c) 4 N   d) None of the above

Answer: (b)
   MC   Moderate

214

16.19: 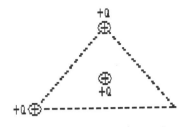 A point charge of +Q is placed at the centroid of an equilateral triangle. When a second charge of +Q is placed at one of the triangle's vertices, an electrostatic force of 4 N acts on it. What is the magnitude of the force that acts on the center charge when a third charge of +Q is placed at one of the other vertices?

    a) Zero   b) 4 N
    c) 8 N   d) None of the above

Answer: (b)
    MC   Difficult

16.20: Two charged objects attract each other with a certain force. If the charges on both objects are doubled with no change in separation, the force between them
    a) quadruples.
    b) doubles.
    c) halves.
    d) increases, but we can't say how much without knowing the distance between them.

Answer: (a)
    MC   Moderate

16.21: Two charged objects attract each other with a force F. What happens to the force between them if one charge is doubled, the other charge is tripled, and the separation distance between their centers is reduced to one-fourth its original value? The force is now equal to

    a) 16 F   b) 24 F   c) 6/16 F   d) 96 F

Answer: (d)
    MC   Moderate

16.22: Consider point charges of +Q and +4Q, which are separated by 3 m. At what point, on a line between the two charges, would it be possible to place a charge of -Q such that the electrostatic force acting on it would be zero?
    a) There is no such point possible.
    b) 1 m from the +Q charge
    c) 1 m from the +4 charge
    d) 3/5 m from the +Q charge

Answer: (b)
    MC   Moderate

16.23: The electric field at the surface of a conductor is always zero.

Answer: False
    TF   Moderate

16.24: Electric field lines
  a) circle clockwise around positive charges.
  b) circle counter-clockwise around positive charges.
  c) radiate outward from negative charges.
  d) radiate outward from positive charges.

Answer: (d)
     MC   Easy

16.25: The electric field shown

  a) increases to the right.   b) increases down.
  c) decreases to the right.   d) decreases down.
  e) is uniform.

Answer: (a)
     MC   Moderate

16.26: A force of 6 N acts on a charge of $3\mu C$ when it is placed in a
   uniform electric field. What is the magnitude of this electric
   field?
  a) 18 MN/C   b) 2 MN/C
  c) 0.5 MN/C   d) None of the above

Answer: (b)
     MC   Easy

16.27: A solid block of metal is placed in a uniform electric field.
   Which statement is correct concerning the electric field in the
   block's interior?
  a) The interior field points in a direction opposite to the
  exterior field.
  b) The interior field points in a direction that is at right
  angles to the exterior field.
  c) The interior points in a direction that is parallel to the
  exterior field.
  d) There is no electric field in the block's interior.

Answer: (d)
     MC   Moderate

16.28: If a solid metal sphere and a hollow metal sphere of equal
   diameters are each given the same charge, the electric field (E)
   midway between the center and the surface is
  a) greater for the solid sphere than for the hollow sphere.
  b) greater for the hollow sphere than for the solid sphere.
  c) zero for both.
  d) equal in magnitude for both, but one is opposite in direction
  from the other.

Answer: (c)
     MC   Moderate

16.29: A hollow metallic sphere is placed in a region permeated by a uniform electric field that is directed upward. Which statement is correct concerning the electric field in the sphere's interior?
   a) The field is zero everywhere in the interior.
   b) The field is directed upward.
   c) The field is directed downward.
   d) The field is zero only at the sphere's exact center.

Answer: (a)
   MC   Difficult

16.30: A positive charge is enclosed in a hollow metallic sphere that is not grounded. At a point directly above the hollow sphere, the electric field caused by the enclosed positive charge has
   a) diminished to zero.   b) diminished somewhat.
   c) increased somewhat.   d) not changed.

Answer: (d)
   MC   Difficult

16.31: A positive point charge is enclosed in a hollow metallic sphere that is grounded. At a point directly above the hollow sphere, the electric field caused by the enclosed positive charge has
   a) diminished to zero.   b) diminished somewhat.
   c) increased somewhat.   d) not changed.

Answer: (a)
   MC   Difficult

16.32: You cannot get an electric shock by touching a charged insulator.

Answer: False
   TF   Easy

16.33: Protons carry a
   a) positive charge.   b) negative charge.
   c) neutral charge.   d) variable charge.

Answer: (a)
   MC   Easy

16.34: Is it possible for two electrons to attract each other?
   a) Yes, they always attract.
   b) Yes, they will attract if they are close enough.
   c) Yes, they will attract if one carries a larger charge than the other.
   d) No, they will never attract.

Answer: (d)
   MC   Easy

16.35: An electron and a proton are separated by a distance of 1 m. What happens to the magnitude of the force on the proton if a second electron is placed next to the first electron?
   a) It quadruples.   b) It doubles.
   c) It will not change.   d) It goes to zero.

Answer: (b)
   MC Easy

16.36: An electron and a proton are separated by a distance of 1 m. What happens to the magnitude of the force on the first electron if a second electron is placed next to the proton?
   a) It doubles.   b) It does not change.
   c) It is reduced to half.   d) It becomes zero.

Answer: (d)
   MC Easy

16.37: An originally neutral electroscope is briefly touched with a positively charged glass rod. The electroscope
   a) remains neutral.
   b) becomes negatively charged.
   c) becomes positively charged.
   d) could become either positively or negatively charged, depending on the time of contact.

Answer: (c)
   MC Easy

16.38: Sphere A carries a net positive charge, and sphere B is neutral. They are placed near each other on an insulated table. Sphere B is briefly touched with a wire that is grounded. Which statement is correct?
   a) Sphere B remains neutral.
   b) Sphere B is now positively charged.
   c) Sphere B is now negatively charged,
   d) The charge on sphere B cannot be determined without additional information.

Answer: (c)
   MC Moderate

16.39: A point charge of +Q is placed at the center of a square. When a second point charge of -Q is placed at one of the square's corners, it is observed that an electrostatic force of 2 N acts on the positive charge at the square's center. Now, identical charges of -Q are placed at the other three corners of the square. What is the magnitude of the net electrostatic force acting on the positive charge at the center of the square?
   a) 0 N   b) 2.8 N   c) 4 N   d) 8 N

Answer: (a)
   MC Moderate

16.40: A large negatively charged object is placed on an insulated table. A neutral metallic ball rolls straight toward the object, but stops before it touches it. A second neutral metallic ball rolls along the path followed by the first ball, strikes the first ball, and stops. The first ball rolls forward, but does not touch the negative object. At no time does either ball touch the negative object. What is the final charge on each ball?
  a) The first ball is positive, and the second ball is negative.
  b) The first ball is negative, and the second ball is positive.
  c) Both balls remain neutral.
  d) Both balls are positive.

Answer: (a)
  MC Difficult

16.41: An originally neutral electroscope is grounded briefly while a positively charged glass rod is held near it. After the glass rod is removed, the electroscope
  a) remains neutral.
  b) is negatively charged.
  c) is positively charged.
  d) could be either positively or negatively charged, depending on how long the contact with ground lasted.

Answer: (b)
  MC Moderate

16.42: Two 0.20 g metal speheres are hung from a common point by nonconducting threads which are 30 cm long. Both are given identical charges, and the electrostatic repulsion forces them apart until the angle between the threads is 20°. How much charge was placed on each sphere?

Answer: 20.4 nC
  ES Difficult

16.43: Three identical charges of 3 μC are placed at the vertices of an equilateral triangle which measures 30 cm on a side. What is the magnitude of the electrostatic force which acts on any one of the charges?

Answer: 1.56 N
  ES Difficult

16.44: Three identical point charges of 2 μC are placed on the x-axis. The first charge is at the origin, the second to the right at x = 50 cm, and the third is at the 100 cm mark. What are the magnitude and direction of the electrostatic force which acts on the charge at the origin?

Answer: 0.18 N left
  ES Moderate

16.45: Three point charges are placed on the x-axis. A charge of +2 μC is placed at the origin, -2 μC to the right at x = 50 cm, and +4 μC at the 100 cm mark. What are the magnitude and direction of the electrostatic force which acts on the charge at the origin?

Answer: 0.072 N left
ES Moderate

16.46: A copper penny has a mass of 3 g. A total of 4×10^{12} electrons are transferred from one neutral penny to another. If the electrostatic force of attraction between the pennies is equal to the weight of a penny, what is the separation between them?

Answer: 35.4 cm
ES Moderate

16.47: At twice the distance from a point charge, the strength of the electric field
a) is four times its original value.
b) is twice its original value.
c) is one-half its original value.
d) is one-fourth its original value.

Answer: (d)
MC Easy

16.48: A solid block of metal is placed in a uniform electric field. Which statement is correct concerning the electric field in the block's interior?
a) The interior field points in a direction opposite to the exterior field.
b) The interior field points in a direction at right angles to the exterior field.
c) The interior field points in a direction which is parallel to the exterior field.
d) There is no electric field in the block's interior.

Answer: (d)
MC Easy

16.49: A cubic block of aluminum rests on a wooden table in a region where a uniform electric field is directed straight upward. What can be said concerning the charge on the block's top surface?
a) The top surface is charged positively.
b) The top surface is charged negatively.
c) The top surface is neutral.
d) The top surface's charge cannot be determined without further information.

Answer: (a)
MC Moderate

16.50: Can electric field lines intersect in free space?
a) Yes, but only at the midpoint between two equal like charges.
b) Yes, but only at the midpoint between a positive and a negative charge.
c) Yes, but only at the centroid of an equilateral triangle with like charges at each corner.
d) No.

Answer: (d)
MC   Moderate

16.51: Consider a square which is 1.0 m on a side. Charges are placed at the corners of the square as follows: +4 μC at (0,0); +4 μC at (1,1); +3 μC at (1,0); -3 μC at (0,1). What is the magnitude of the electric field at the square's center?

Answer: 1.08 x 10^5 N/C
ES   Difficult

16.52: A metal sphere of radius 10 cm carries a charge of +2 μC. What is the magnitude of the electric field 5 cm from the sphere's surface?

Answer: 8 x 10^5 N/C
ES   Moderate

16.53: How can a negatively charged rod charge an electroscope positively?
a) By conduction.   b) By induction.
c) By deduction.   d) It cannot.

Answer: (b)
MC   Easy

16.54: How can a negatively charged rod charge an electroscope negatively?
a) By conduction.   b) By induction.
c) By deduction.   d) It cannot.

Answer: (a)
MC   Easy

16.55: What is the force between a 1.5 μC charge and a 2.5 μC charge which are 15 cm apart?
a) 13.5 N   b) 10.0 N   c) 1.5 N   d) 0.225 N

Answer: (c)
MC   Easy

16.56: Two 1 C charges have a force between them of 1 N. How far apart
are they?
   a) 95 km   b) 9.5 m   c) 4 m   d) 4 mm

Answer: (a)
   MC   Easy

16.57: The force between a 0.015 C charge and a 10.3μC charge is 0.4 N.
How far apart are they?
   a) 59 m   b) 39 m   c) 39 cm   d) 59 mm

Answer: (a)
   MC   Easy

16.58: A 1 C charge is 15 m from a second charge, and the force between
them is 1 N. What is the magnitude of the second charge?
   a) 25 C   b) 1 C   c) 0.025 C   d) 25 nC

Answer: (d)
   MC   Easy

16.59: An atom has more electrons than protons. The atom is
   a) a positive ion.   b) a negative ion.
   c) a superconductor.   d) impossible.

Answer: (b)
   MC   Easy

16.60: A positive object touches a neutral electroscope, and the leaves
separate. Then a negative object is brought near the
electroscope, but does not touch it. What happens to the leaves?
   a) They separate further.
   b) They move closer together.
   c) They are unaffected.
   d) Cannot be determined without further information.

Answer: (b)
   MC   Moderate

16.61: What is the charge on 1 kg of protons?
   a) 1 C   b) 1000 C
   c) 9.6 x 10^7 C   d) 6.0 x 10^{26} C

Answer: (c)
   MC   Moderate

16.62: An atomic nucleus has a charge of +40e. An electron is 10^{-9}m
from the nucleus. What is the force on the electron?

Answer: 9.2 nN
   ES   Easy

16.63: A +30 μC charge is attracted to a -90 μC with a force of 1.8 N. How far apart are the charges?

Answer: 3.7 m
ES   Easy

16.64: An electron is held up against the force of gravity by the attraction of a fixed proton some distance above it. How far above the electron is the proton?

Answer: 5.08 m
ES   Moderate

16.65: A styrofoam ball of mass 0.12 g is placed in an electric field of 6000 N/C pointing downward. What charge must be placed on the ball for it to be suspended?

Answer: -196 nC
ES   Moderate

16.66: A foam ball of mass 0.15 g carries a charge of -2 nC. The ball is placed inside a uniform electric field, and is suspended against the force of gravity. What are the magnitude and direction of the electric field?

Answer: 735 kN/C downward
ES   Moderate

16.67: An atomic nucleus has a charge of +40e. What is the magnitude of the electric field at a distance of 1 m from the nucleus?

Answer: 5.76 x 10^{10} N/C
ES   Easy

16.68: What are the magnitude and direction of the electric field at a distance of 1.2 m from a -20 nC charge?

Answer: 125 N/C toward the charge
ES   Easy

17. Electric Potential and Electric Energy; Capacitance

17.1: **MATCH THE UNIT TO THE PHYSICAL QUANTITY.**

17.1: dielectric constant

Answer: dimensionless
MA   Easy

17.2: electric potential

Answer: volt
MA   Easy

17.3: capacitance

Answer: farad
MA   Easy

17.4: electric potential energy

Answer: electron-volt
MA   Easy

17.5: Which of the following is an accurate statement?
a) All parts of a perfect conductor are at the same potential.
b) If a solid metal sphere carries a net charge, the charge will be uniformly distributed throughout the volume of the sphere.
c) A conductor cannot carry a net charge.
d) The electric field at the surface of a conductor is not necessarily perpendicular to the surface in all cases.

Answer: (a)
MC   Easy

17.6: How much energy is necessary to place three charges, each of $2\mu C$, at the corners of an equilateral triangle of side 2 cm?

Answer: 5.4 J
ES   Easy

17.7: Equipotential lines always begin and end on charges.

Answer: False
TF   Easy

17.8:

The net work done in moving an electron from point A at -50 V to point B at +50 V along the semi-circular path shown is
a) $+1.6 \times 10^{-17}$ J
b) -1.6×10^{-17} J
c) zero
d) cannot be determined; not enough information given.

Answer: (b)
MC   Moderate

17.9: Starting from rest, a proton falls through a potential difference of 1200 V. What speed does it acquire?

Answer: 4.8×10^5 m/s
ES   Moderate

17.10: Four charges of equal charge +q are placed at the corners of a rectangle of sides a and b. What is the potential at the center of the rectangle if q = 2 μC, a = 3 cm, and b = 4 cm?

Answer: 2.9×10^6 V
ES   Moderate

17.11: A metal sphere of radius 8 cm is charged to a potential of -500 V. With what velocity must an electron be fired toward the sphere if it is to just barely reach the sphere when started from a position 15 cm from the center of the sphere?

Answer: 9×10^6 m/s
ES   Difficult

17.12: It takes 10 J of energy to move 2 C of charge from point A to point B. What is the potential difference between points A and B?
a) 20 V
b) 0.2 V
c) 5 V
d) None of the above

Answer: (c)
MC   Easy

17.13: If a Cu^{2+} ion drops through a potential difference of 12 V, it will acquire a kinetic energy (in the absence of friction) of
a) 12 eV
b) 6 eV
c) 24 eV
d) none of the above.

Answer: (c)
MC   Easy

17.14: The electron-volt is a unit of
   a) voltage.   b) current.   c) power.   d) energy.

Answer: (d)
   MC   Easy

17.15: Consider a uniform electric field of 50 N/C directed toward the east. If the voltage measured relative to ground at a given point in the field is 80 V, what is the voltage at a point 1 m directly east of the point?
   a) 15 V
   b) 30 V
   c) 130 V
   d) Impossible to calculate from the information given.

Answer: (b)
   MC   Moderate

17.16: Consider a uniform electric field of 50 N/C directed toward the east. If the voltage measured relative to ground at a given point in the field is 80 V, what is the voltage at a point 1 m directly south of that point?
   a) zero   b) 30 V   c) 50 V   d) 80 V

Answer: (d)
   MC   Moderate

17.17: The absolute potential at a distance of 2 m from a positive point charge is 100 V. What is the absolute potential 4 m away from the same point charge?
   a) 25 V .   b) 50 V   c) 200 V   d) 400 V

Answer: (b)
   MC   Moderate

17.18: The absolute potential at a distance of 2 m from a negative point charge is -100 V. What is the absolute potential 4 m away from the same point charge?
   a) -25 V   b) -50 V   c) -200 V   d) -400 V

Answer: (b)
   MC   Moderate

17.19: The absolute potential at the center of a square is 3 V when a charge of +Q is located at one of the square's corners. What is the absolute potential at the square's center when a second charge of -Q is placed at one of the remaining corners?
   a) Zero   b) 3 V
   c) 6 V   d) None of the above

Answer: (a)
   MC   Moderate

17.20: Which of the following is a vector?
   a) Electric potential   b) Electric charge
   c) Electric field   d) All of the above

Answer: (c)
   MC   Easy

17.21: Several electrons are placed on a hollow conducting sphere. They
   a) clump together on the sphere's outer surface.
   b) clump together on the sphere's inner surface.
   c) become uniformly distributed on the sphere's outer surface.
   d) become uniformly distributed on the sphere's inner surface.

Answer: (c)
   MC   Moderate

17.22: What charge appears on the plates of a 2 μF capacitor when it is
   charged to 100 V?

Answer: 200 μC
   ES   Easy

17.23: A parallel plate capacitor is constructed with plate area of 0.4
   m^2 and a plate separation of 0.1 mm. (a) How much charge is
   stored on it when it is charged to a potential difference of 12
   V?   (b) How much energy is stored?

Answer: (a) 0.42 μC   (b) 2.5 μJ
   ES   Moderate

17.24: Electric dipoles always consist of two charges that are
   a) equal in magnitude; opposite in sign.
   b) equal in magnitude; both are negative.
   c) equal in magnitude; both are positive.
   d) unequal in magnitude; opposite in sign.

Answer: (a)
   MC   Easy

17.25: A battery charges a parallel-plate capacitor fully and then is
   removed. The plates are immediately pulled apart. (With the
   battery disconnected, the amount of charge on the plates remains
   constant.) What happens to the potential difference between the
   plates as they are being separated?
   a) It increases.
   b) It decreases.
   c) It remains constant.
   d) There is no way to tell from the information given.

Answer: (a)
   MC   Moderate

17.26: The plates of a parallel-plate capacitor are maintained with constant voltage by a battery as they are pulled apart. During this process, the amount of charge on the plates must
a) increase.
b) decrease.
c) remain constant.
d) either increase or decrease. There is no way to tell from the information given.

Answer: (b)
MC   Moderate

17.27: The plates of a parallel-plate capacitor are maintained with constant voltage by a battery as they are pulled apart. What happens to the strength of the electric field during this process?
a) It increases.
b) It decreases.
c) It remains constant.
d) There is no way to tell from the information given.

Answer: (b)
MC   Moderate

17.28: A parallel-plate capacitor has plates of area 0.2 m^2 separated by a distance of 0.001 m. What is the strength of the electric field between these plates when this capacitor is connected to a 6-V battery?
a) 1200 N/C   b) 3000 N/C
c) 6000 N/C   d) None of the above

Answer: (c)
MC   Moderate

17.29: A parallel-plate capacitor has plates of area 0.2 m^2 separated by a distance of 0.001 m. What is this capacitor's capacitance?
a) 200 F   b) 40 F
c) 1.77×10^{-9} F   d) 3.54×10^{-10} F

Answer: (c)
MC   Easy

17.30: A charge of 2 μC flows onto the plates of a capacitor when it is connected to a 12-V battery. How much work was done in charging this capacitor?
a) 24 μJ   b) 12 μJ
c) 144 μJ   d) None of the above

Answer: (b)
MC   Easy

17.31: A 15-μF capacitor is connected to a 50-V battery and becomes fully charged. The battery is removed and a slab of dielectric that completely fills the space between the plates is inserted. If the dielectric has a dielectric constant of 5, what is the capacitance of the capacitor after the slab is inserted?
   a) 75 μF
   b) 20 μF
   c) 3 μF
   d) None of the above

Answer: (a)
   MC   Moderate

17.32: A 15-μF capacitor is connected to a 50-V battery and becomes fully charged. The battery is removed and a slab of dielectric that completely fills the space between the plates is inserted. If the dielectric has a dielectric constant of 5, what is the voltage across the capacitor's plates after the slab is inserted?
   a) 250 V
   b) 10 V
   c) 2 V
   d) None of the above

Answer: (b)
   MC   Moderate

17.33: All parts of a perfect conductor are at the same potential.

Answer: True
   TF   Easy

17.34: If a solid metal sphere carries a net charge, the charge will be uniformly distributed throughout the volume of the sphere.

Answer: False
   TF   Easy

17.35: A conductor cannot carry a net charge.

Answer: False
   TF   Easy

17.36: The electric field at the surface of a conductor is always perpendicular to the surface.

Answer: True
   TF   Easy

17.37: The electric field at the surface of a conductor is always zero.

Answer: False
   TF   Easy

17.38: Consider a uniform electric field of 50 N/C directed toward the east. If the voltage measured relative to ground at a given point is 80 V, what is the voltage at a point 1 m directly west of that point?
a) 30 V
b) 80 V
c) 130 V
d) Impossible to calculate from the information given.

Answer: (c)
MC   Moderate

17.39: The absolute potential at the exact center of a square is 3 V when a charge of +Q is located at one of the square's corners. What is the absolute potential at the square's center when each of the other corners is also filled with a charge of +Q?
a) 0 V   b) 3 V   c) 9 V   d) 12 V

Answer: (d)
MC   Moderate

17.40: What is the electric potential infinitely far away from a point charge of +Q?
a) $+9 \times 10^9$ V   b) -9×10^9 V   c) 0 V   d) infinite

Answer: (c)
MC   Easy

17.41: A square is 1 m on a side. Charges of +4 μC are placed in two diagonally opposite corners. In the other two corners are placed charges of +3 μC and -3 μC. What is the absolute potential in the square's center?

Answer: 1.02×10^5 V
ES   Moderate

17.42: A stationary electron is accelerated through a potential difference of 500 V. What is the velocity of the electron afterward?

Answer: 1.33×10^7 m/s
ES   Moderate

17.2: The diagonals of a given square are 2 m long. This square has a +6 μC charge in one corner, and a +5 μC charge in an adjacent corner. Refer to this square for the next three problems.

17.43: What is the potential difference between the vacant corner adjacent to the +6 μC and the square's center?

Answer: 3.83×10^4 V
ES   Moderate   Table: 2

17.44: What is the potential difference between the two vacant corners?

Answer: 1860 V
ES   Moderate   Table: 2

17.45: How much energy is required to move an electron from the square's center to a position infinitely far away?

Answer: 1.59×10^{-14} J
ES   Moderate   Table: 2

17.46: Two parallel-plate capacitors are identical in every respect except that one has twice the plate area of the other. If the smaller capacitor has capacitance C, the larger one has capacitance
   a) C/2   b) C   c) 2C   d) 4C

Answer: (c)
MC   Easy

17.47: If the electric field between the plates of a given capacitor is weakened, the capacitance of that capacitor
   a) increases.
   b) decreases.
   c) does not change.
   d) cannot be determined from the information given.

Answer: (c)
MC   Easy

17.48: A dielectric material such as paper is placed between the plates of a capacitor holding a fixed charge. What happens to the electric field between the plates?
   a) No change.   b) Becomes stronger.
   c) Becomes weaker.   d) Reduces to zero.

Answer: (c)
MC   Moderate

17.49: A dielectric material such as paper is placed between the plates of a capacitor. What happens to the capacitance?
   a) No change.   b) Becomes larger.
   c) Becomes smaller.   d) Becomes infinite.

Answer: (b)
MC   Moderate

17.50: A parallel-plate capacitor is connected to a battery and becomes fully charged. The capacitor is then disconnected, and the separation between the plates is increased in such a way that no charge leaks off. The energy stored in this capacitor has
a) increased.
b) decreased.
c) not changed.
d) become zero.

Answer: (a)
MC   Moderate

17.51: Doubling the voltage across a given capacitor causes the energy stored in that capacitor to
a) quadruple.
b) double.
c) reduce to one half.
d) reduce to one fourth.

Answer: (a)
MC   Easy

17.52: Doubling the capacitance of a capacitor holding a constant charge causes the energy stored in that capacitor to
a) quadruple.
b) double.
c) decrease to one half.
d) decrease to one fourth.

Answer: (d)
MC   Easy

17.53: A 6 μF air capacitor is connected across a 100 V battery. After the battery fully charges the capacitor, the capacitor is immersed in transformer oil (dielectric constant = 4.5). How much additional charge flows from the battery, which remained connected during the process?

Answer: 2.1 mC
ES   Moderate

17.54: What is the potential difference if 0.58 J of work is needed to move 0.06 C of charge?
a) 6.3 V   b) 0.1 V   c) 0.03 V   d) 9.7 V

Answer: (d)
MC   Easy

17.55: How much work does 9.0 V do in moving 8.5 x 10^{18} electrons?
a) 12.24 J   b) 7.65 J   c) 1.36 J   d) 1.06 J

Answer: (a)
MC   Moderate

17.56: Two identical aluminum objects are insulated from their
surroundings. Object A has a net charge of excess electrons.
Object B is grounded. Which object is at a higher potential?
  a) A
  b) B
  c) Both are at the same potential.
  d) Cannot be determined without more information.

Answer: (b)
      MC   Moderate

17.57: A 4 g object carries a charge of 20 μC. The object is
accelerated from rest through a potential difference, and
afterward the ball is moving at 2.0 m/s. What is the magnitude
of the potential difference?
  a) 800 kV   b) 400 kV   c) 800 V   d) 400 V

Answer: (d)
      MC   Moderate

17.58: A small charged ball is accelerated from rest to a speed v by a
500 V potential difference. If the potential difference is
changed to 2000 V, what will the new speed of the ball be?
  a) v   b) 2 v   c) 4 v   d) 16 v

Answer: (b)
      MC   Moderate

17.59: A parallel plate capacitor has a plate separation of 5 cm. If
the potential difference between the plates is 2000 V, with the
top plate at the higher potential, what is the electric field
beteeen the plates?
  a) 100 N/C upward   b) 100 N/C downward
  c) 40000 N/C upward   d) 40000 N/C downward

Answer: (d)
      MC   Moderate

17.60: What is the potential at a distance of 5×10^{-10} m from a nucleus
of charge +60e?

Answer: 173 V
      ES   Moderate

17.61: An alpha particle (charge +2e, mass 6.64×10^{-27}) moves head-on
at a fixed gold nucleus (charge +79e). If the distance of
closest approach is 2×10^{-10} m, what was the initial speed of
the alpha particle?

Answer: 2.34×10^5 m/s
      ES   Difficult

17.62: A parallel-plate capacitor is filled with air, and the plates are separated by 0.05 mm. If the capacitance is 17.3 pF, what is the plate area?

Answer: 9.77×10^{-5} m^2
ES   Moderate

17.63: 20 V is placed across a 15 μF capacitor. What is the energy stored in the capacitor?
a) 150 μJ   b) 300 μJ   c) 3 mJ   d) 6 mJ

Answer: (c)
MC   Easy

17.64: A charge of 60 μC is placed on a 15 μF capacitor. How much energy is stored in the capacitor?
a) 120 J   b) 4 J   c) 240 μJ   d) 120 μJ

Answer: (d)
MC   Easy

18. Electric Currents

18.1: MATCH THE UNIT TO THE PHYSICAL QUANTITY.

18.1: electric resistance

Answer: ohm
      MA   Easy

18.2: temperature coefficient of resistivity

Answer: $(°C)^{-1}$
      MA   Easy

18.3: electric current

Answer: ampere
      MA   Easy

18.4: resistivity

Answer: Ω-m
      MA   Easy

18.5: electric power

Answer: volt-amp
      MA   Easy

18.6: A car battery
      a) has an emf of 6 V consisting of one 6-V cell.
      b) has an emf of 6 V consisting of three 2-V cells connected in series.
      c) has an emf of 6 V consisting of three 2-V cells connected in parallel.
      d) has an emf of 12 V consisting of six 2-V cells connected in series.
      e) has an emf of 12 V consisting of six 2-V cells connected in parallel.

Answer: (d)
      MC   Moderate

18.7: If you connect two identical storage batteries together in parallel, and place them in a circuit, the combination will provide
  a) twice the voltage and twice the total charge that one battery would.
  b) twice the voltage and the same total charge that one battery would.
  c) the same voltage and twice the total charge that one battery would.
  d) half the voltage and half the total charge that one battery would.
  e) half the voltage and twice the total charge that one battery would.

Answer: (c)
  MC Moderate

18.8: If you connect two identical storage batteries together in series ("+" to "-" to "+" to "-"), and place them in a circuit, the combination will provide
  a) zero volts.
  b) twice the voltage, and different currents will flow through each.
  c) twice the voltage, and the same current will flow through each.
  d) the same voltage, and different currents will flow through each.
  e) the same voltage and the same current will flow through each.

Answer: (c)
  MC Moderate

18.9: Car batteries are rated in "amp-hours." This is a measure of their
  a) charge.   b) current.   c) emf.   d) power.

Answer: (a)
  MC Easy

18.10: The net direction in which electrons flow through a circuit is called conventional current.

Answer: False
  TF Easy

18.11: A charge of 12 C passes through an electroplating apparatus in 2 min. What is the average current?

Answer: 0.1 A
  ES Easy

236

18.12: A battery is rated at 12 V and 160 A-h. How much energy does the battery store?

Answer: 6.9 MJ
ES Moderate

18.13: The diameter of no. 12 copper wire is 0.081 in. The maximum safe current it can carry (in order to prevent fire danger in building construction) is 20 A. At this current, what is the drift velocity of the electrons? (The number of electron carriers in one cubic centimeter of copper is 8.5×10^{22}.)

Answer: 0.44 mm/s
ES Difficult

18.14: In an electroplating process, it is desired to deposit 40 mg of silver on a metal part by using a current of 2 A. How long must the current be allowed to run to deposit this much silver? (The silver ions are singly charged, and the atomic weight of silver is 108.)

Answer: 17.8 s
ES Moderate

18.15: A coulomb per second is the same as
   a) a watt.   b) an ampere.
   c) a volt-second.   d) a volt per second.

Answer: (b)
MC Easy

18.16: Consider two copper wires each carrying a current of 3 A. One wire has twice the diameter of the other. The ratio of the drift velocity in the smaller diameter wire to that in the larger diameter wire is
   a) 4:1   b) 2:1   c) 1:2   d) 1:4

Answer: (a)
MC Moderate

18.17: A coffee maker, which draws 13.5 A of current, has been left on for 10 min. What is the net number of electrons that have passed through the coffee maker?

Answer: 5.06×10^{22}
ES Moderate

18.18: A device obeying Ohm's law must have a constant resistance.

Answer: True
TF Easy

18.19: A potential difference (V) is applied to the ends of an object causing a current (I) to flow through it. The object obeys Ohm's law because we can obtain the resistance by dividing the voltage by the current: R = V/I.

Answer: False
      TF Moderate

18.20: A lightbulb is an example of a device that obeys Ohm's law.

Answer: False
      TF Moderate

18.21: What potential difference is required to cause 2 A to flow through a resistance of 8 Ω?

Answer: 16 V
      ES Moderate

18.22: What is the resistance of 1.0 m of no. 18 copper wire (diameter 0.40 in)? (The resistivity of copper is 1.69×10^{-8} Ω-m.)

Answer: 0.00021 Ω
      ES Moderate

18.23: A heavy bus bar is 20 cm long and of rectangular cross-section, 1 cm x 2 cm. What is the voltage drop along its length when it carries 4000 A? (The resistivity of copper is $1.69 \times 10-8$ Ω-m.)

Answer: 0.068 V
      ES Easy

18.24: The temperature coefficient of resistivity of platinum is $3.9 \times 10^{-3}/°K$. If a platinum wire has a resistance of R at room temperature (23°C), to what temperature must it be heated in order to double its resistance to 2R?

Answer: 279°C
      ES Moderate

18.25: The resistance of a wire is defined as
    a) (length)(resistivity)/(cross-sectional area).
    b) (current)/(voltage).
    c) (voltage)/(current).
    d) none of the above.

Answer: (c)
      MC Moderate

18.26: Consider two copper wires. One has twice the length and twice the cross-sectional area of the other. How do the resistances of these two wires compare?
a) Both wires have the same resistance.
b) The longer wire has twice the resistance of the shorter wire.
c) The longer wire has four times the resistance of the shorter wire.
d) None of the above.

Answer: (a)
MC Moderate

18.27: Consider two copper wires. One has twice the length of the other. How do the resistivities of these two wires compare?
a) Both wires have the same resistivity.
b) The longer wire has twice the resistivity of the shorter wire.
c) The longer wire has four times the resistivity of the shorter wire.
d) None of the above.

Answer: (a)
MC Easy

18.28: Negative temperature coefficients of resistivity
a) do not exist.   b) exist in conductors.
c) exist in semiconductors.   d) exist in superconductors.

Answer: (c)
MC Easy

18.29: Which of the following graphs indicates the behavior of a superconductor?

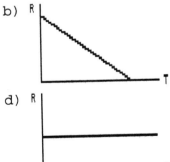

Answer: (c)
MC Easy

18.30: Which of the following graphs indicates that the material is a semiconductor?

a)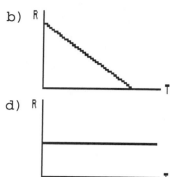

b)

c)

d)

Answer: (b)
       MC   Easy

18.31: Which of the following graphs indicates that Ohm's law is obeyed for the range shown?

a)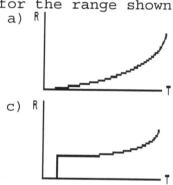

b)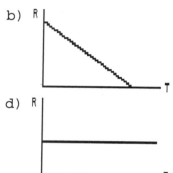

c)

d)

Answer: (d)
       MC   Easy

18.32: If the resistance in a constant voltage circuit is doubled, the power dissipated by that circuit will
    a) increase by a factor of two.
    b) increase by a factor of four.
    c) decrease to one-half its original value.
    d) decrease to one-fourth its original value.

Answer: (c)
       MC   Moderate

18.33: The resistivity of most common metals
    a) remains constant over wide temperature ranges.
    b) increases as the temperature increases.
    c) decreases as the temperature increases.
    d) varies randomly as the temperature increases.

Answer: (b)
       MC   Easy

18.34: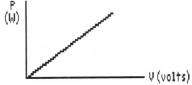

In the graph shown, what physical quantity does the slope represent?
a) Current
b) Energy
c) Resistivity
d) Resistance

Answer: (a)
MC   Moderate

18.35: A 500-W device is connected to a 100-V power source. What current flows through this surface?
a) 50,000 A
b) 0.2 A
c) 5 A
d) None of the above

Answer: (c)
MC   Easy

18.36: A lamp uses a 150-W bulb. If it is used at 120 V, what current does it draw and what is its resistance?

Answer: 1.25 A; 96 Ω
ES   Moderate

18.37: A 200-Ω resistor is rated at 1/4 W. (a) What is the maximum current it can draw? (b) What is the maximum voltage that should be applied across it?

Answer: (a) 0.0354 A   (b) 7.07 V
ES   Moderate

18.38: A motor that can do work at the rate of 2 hp has 60% efficiency. How much current does it draw from a 120 V line? (1 hp = 746 W.)

Answer: 20.7 A
ES   Moderate

18.39: The monthly (30 days) electric bill included the cost of running a central air-conditioning unit for 2 hr/day at 5000 W, and a series connection of ten 60-W lightbulbs for 5 hr/day. How much did these items contribute to the cost of the monthly electric bill if electricity costs 8¢ per kWh?

Answer: $31.20
ES   Moderate

18.40: The heating element in an electric drier operates on 240 V and generates heat at the rate of 2 kW. The heating element shorts out and, in repairing it, the owner shortens the Nichrome wire by 10%. (Assume the temperature is unchanged. In reality, the resistivity of the wire will depend on its temperature.) What effect will the repair have on the power dissipated in the heating element?

Answer: Power dissipated increases to 2.22 kW
ES   Moderate

18.41: Appliances in the USA are designed to work on 110-120 V, whereas appliances in Europe operate on 220-240 V. What would happen if you tried to use a European electric shaver in this country?
a) It would probably work as well as it does in Europe, since it is the power rating, not the voltage rating, that matters.
b) It would probably work as well as it does in Europe, since it is the current rating, not the voltage rating, that matters.
c) It would barely work (or maybe not work at all), since the voltage here is too low to push enough current through the device.
d) It would probably overheat and burn up before very long.
e) This question cannot be answered without knowing the frequency used in Europe and in the USA , since this is the parameter that matters most in any electrical device that operates on alternating current.

Answer: (c)
MC   Moderate

18.42: If the voltage across a circuit of constant resistance is doubled, the power dissipated by that circuit will
a) quadruple.   b) double.
c) decrease to one half.   d) decrease to one fourth.

Answer: (a)
MC   Easy

18.43: If the current flowing through a circuit of constant resistance is doubled, the power dissipated by that circuit will
a) quadruple.   b) double.
c) decrease to one half.   d) decrease to one fourth.

Answer: (a)
MC   Easy

18.44: A 1 mm diameter copper wire (resistivity 1.69×10^{-8} $\Omega \cdot m$) carries a current of 15 A. What is the potential difference between two points 100 m apart?

Answer: 32.3 V
ES   Moderate

18.45: A total of 2 x 10^{13} protons pass a given point in 15 s. What is the current?

Answer: 0.21 μA
        ES Moderate

18.46: A toaster is rated 800 W at 120 V. What is the resistance of its heating element?

Answer: 18 Ω
        ES Easy

18.47: What current is flowing if 0.47 C of charge pass a point in 0.2 s?
    a) 2.35 A   b) 0.47 A   c) 0.094 A   d) 0.2 A

Answer: (a)
        MC Easy

18.48: What current is flowing if 4 x 10^{16} electrons pass a point in 0.5 s?
    a) 0.013 A   b) 0.312 A   c) 6.25 A   d) 78.13 A

Answer: (a)
        MC Moderate

18.49: How much charge must pass by a point in 1.5 s for the current to be 2.0 A?
    a) 0.75 C   b) 1.0 C   c) 2.0 C   d) 3.0 C

Answer: (d)
        MC Easy

18.50: What is the resistance of a circular rod 1 cm in diameter and 45 m long, if the resistivity is 1.4 x 10^{-8} $\Omega \cdot$m?
    a) 0.0063 Ω b) 0.008 Ω c) 0.8 Ω d) 6.3 Ω

Answer: (b)
        MC Moderate

18.51: A 1.5 cm square rod, 4 m long, measures 0.04 ohms. What is its resistivity?
    a) 0.0225 $\Omega \cdot$m   b) 0.015 $\Omega \cdot$m
    c) 1.5 x 10^{-4} $\Omega \cdot$m   d) 2.25 x 10^{-6} $\Omega \cdot$m

Answer: (d)
        MC Moderate

18.52: What length of copper wire (resistivity 1.7×10^{-8} $\Omega \cdot$m) of diameter 0.15 mm is needed for a total resistance of 15 Ω?
   a) 15.6 mm   b) 15.6 cm   c) 1.56 m   d) 15.6 m

Answer: (d)
   MC   Moderate

18.53: How much more resistance does a 1 cm diameter rod have compared to a 2 cm diameter rod of the same length and made of the same material?
   a) 75%   b) 100%   c) 300%   d) 400%

Answer: (c)
   MC   Moderate

18.54: A light bulb operating at 110 V draws 1.4 A of current. What is its resistance?
   a) 12.7 Ω b) 78.6 Ω c) 109 Ω d) 154 Ω

Answer: (b)
   MC   Easy

18.55: A resistance of 330 Ω has 4 A flowing through it. What is the voltage across the resistance?
   a) 12.1 V   b) 82.5 V   c) 334 V   d) 1320 V

Answer: (d)
   MC   Easy

18.56: A 4000 Ω resistor is connected across 220 V. What current will flow?
   a) 0.06 A   b) 1.8 A   c) 5.5 A   d) 18.2 A

Answer: (a)
   MC   Easy

18.57: A 110 V hairdryer is rated at 1200 W. What current will it draw?
   a) 0.09 A   b) 1.0 A   c) 10.9 A   d) 12.0 A

Answer: (c)
   MC   Easy

18.58: 4.0 A is flowing through an 8.0 Ω resistor. How much power is being dissipared?
   a) 32 W   b) 128 W   c) 256 W   d) 1024 W

Answer: (b)
   MC   Easy

18.59: A 25 W soldering iron runs on 110 V. What is its resistance?
a) 0.002 Ω   b) 4.4 Ω   c) 484 Ω   d) 2750 Ω

Answer: (c)
MC   Easy

18.60: A 150 W light bulb running on 110 V draws how much current?
a) 0.73 A   b) 1.4 A   c) 2.0 A   d) 15 A

Answer: (b)
MC   Easy

18.61: How much energy does a 100 W light bulb use in 8 hours?
a) 0.008 kWh   b) 0.8 kWh   c) 12.5 kWh   d) 800 kWh

Answer: (b)
MC   Easy

18.62: How much energy does a 25 W soldering iron use in 8 hours?
a) 400 J   b) 11250 J   c) 12000 J   d) 720,000 J

Answer: (d)
MC   Moderate

18.63: How much does it cost to operate a 25 W soldering iron for 8 hours, if energy costs $0.08/kWh?
a) $1.50   b) $0.25   c) $0.16   d) $0.016

Answer: (d)
MC   Moderate

18.64: 14 A of current flows through 8.0 Ω for 24 hours. How much does this cost if energy costs $0.09/kWh?
a) $0.24   b) $1.04   c) $2.16   d) $3.39

Answer: (d)
MC   Moderate

18.65: Resistivity is measured in ohms.

Answer: False
TF   Easy

18.66: Electric charge past a point per unit time is current.

Answer: True
TF   Easy

18.67: Energy per unit charge is power.

Answer: False
       TF   Easy

18.68: Resistance is the opposition to the flow of charge.

Answer: True
       TF   Easy

18.69: What is 1 Ω equivalent to?
       a) 1 J/s   b) 1 W/A   c) 1 V·A   d) 1 V/A

Answer: (d)
       MC   Moderate

18.70: What is 1 W equivalent to?
       a) 1 V/A   b) 1 Ω·A   c) 1 V·A   d) 1 V/Ω

Answer: (c)
       MC   Moderate

18.71: A kilowatt-hour is equivalent to
       a) 1000 W   b) 3600 s
       c) 3,600,000 J/s   d) 3,600,000 J

Answer: (d)
       MC   Easy

18.72: A 9 V battery costs $1.49, and will run a portable CD player for 6 hours. Suppose the battery supplies a current of 25 mA to the player. What is the cost of energy in dollars per kWh?

Answer: $1104/kWh
       ES   Difficult

18.73: A 120 m long copper wire (resistivity 1.7×10^{-8} Ω·m) has resistance 6.0 Ω. What is the diameter of the wire?

Answer: 0.66 mm
       ES   Difficult

18.1: **MATCH THE UNIT TO THE PHYSICAL QUANTITY.**
      TA

19. DC Circuits and Instruments

19.1: **MATCH THE DEVICE TO THE ELECTRICAL SYMBOL.**

19.1:

Answer: ground
MA   Easy

19.2:

Answer: voltmeter
MA   Easy

19.3:

Answer: resistor
MA   Easy

19.4:

Answer: capacitor
MA   Easy

19.5:

Answer: battery
MA   Easy

19.6:

Answer: galvanometer
MA   Easy

19.7:

Answer: switch
       MA   Easy

19.8:

Answer: ammeter
       MA   Easy

19.9: When two or more resistors are connected in series to a battery
       a) the total voltage across the combination is the algebraic sum
       of the voltages across the individual resistors.
       b) the same current flows through each resistor.
       c) the equivalent resistance of the combination is equal to the
       sum of the resistances of each resistor.
       d) all of the above.

Answer: (d)
       MC   Easy

19.10: When two or more resistors are connected in parallel to a
       battery,
       a) the voltage across each resistor is the same.
       b) the total current flowing from the battery equals the sum of
       the currents flowing through each resistor.
       c) the equivalent resistance of the combination is less than the
       resistance of any one of the resistors.
       d) all of the above.

Answer: (d)
       MC   Easy

19.11: Three identical resistors are connected in series to a battery.
       If the current of 12 A flows from the battery, how much current
       flows through any one of the resistors?
       a) 12 A   b) 4 A
       c) 36 A   d) None of the above

Answer: (a)
       MC   Easy

19.12: Three identical resistors are connected in parallel to a battery. If the current of 12 A flows from the battery, how much current flows through any one of the resistors?
a) 12 A
b) 4 A
c) 36 A
d) None of the above

Answer: (b)
MC   Easy

19.13: Three identical resistors are connected in series to a 12-V battery. What is the voltage across any one of the resistors?
a) 36 V
b) 12 V
c) 4 V
d) None of the above

Answer: (c)
MC   Easy

19.14: Three identical resistors are connected in parallel to a 12-V battery. What is the voltage of any one of the resistors?
a) 36 V
b) 12 V
c) 4 V
d) None of the above

Answer: (b)
MC   Easy

19.15: A 6-Ω and a 12-Ω resistor are connected in series to a 36-V battery. What power is dissipated by the 6-Ω resistor?
a) 216 W
b) 48 W
c) 486 W
d) None of the above

Answer: (d)
MC   Moderate

19.16: A 6-Ω and a 12-Ω resistor are connected in parallel to a 36-V battery. What power is dissipated by the 6-Ω resistor?
a) 216 W
b) 48 W
c) 486 W
d) None of the above

Answer: (a)
MC   Moderate

19.17: As more resistors are added in series to a constant voltage source, the power supplied by the source
a) increases.
b) decreases.
c) does not change.
d) increases for a time and then starts to decrease.

Answer: (b)
MC   Moderate

19.18: As more resistors are added in parallel to a constant voltage
       source, the power supplied by the source
       a) increases.
       b) decreases.
       c) does not change.
       d) increases for a time and then starts to decrease.

Answer: (a)
        MC Moderate

19.19: 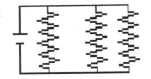 If identical resistors are connected in
       parallel, and a battery
       is connected as shown, the resistor closest
       to the battery will have the largest current
       flowing through it.

Answer: False
        TF Easy

19.20: You obtain a 100-W lightbulb and a 50-W light bulb. Instead of
       connecting them in the normal way, you devise a circuit that
       places them in series across normal household voltage. Which
       statement is correct?
       a) Both bulbs glow at the same reduced brightness.
       b) Both bulbs glow at the same increased brightness.
       c) The 100-W bulb glows brighter than the 50-W bulb.
       d) The 50-W bulb glows more brightly than the 100-W bulb.

Answer: (d)
        MC Difficult

19.21: The lamps in a string of Christmas tree lights are connected in
       parallel. What happens if one lamp burns out? (Assume
       negligible resistance in the wires leading to the lamps.)
       a) The brightness of the lamps will not change appreciably.
       b) The other lamps get brighter equally.
       c) The other lamps get brighter, but some get brighter than
       others.
       d) The other lamps get dimmer equally.
       e) The other lamps get dimmer, but some get dimmer than others.

Answer: (a)
        MC Moderate

19.22: Four 20-Ω resistors are connected in parallel. What is the
       equivalent resistance?

Answer: 5 Ω
        ES Easy

19.23: A combination of 2 Ω in series with 4 Ω is connected in parallel with 3 Ω. What is the equivalent resistance?

Answer: 2 Ω
ES   Moderate

19.24: Two 100-W lightbulbs are to be connected to a 120 V source. What is the current, potential difference, and dissipated power for each when they are connected (a) in parallel (the normal arrangement)? (b) in series?

Answer: (a) 0.83 A in each; 120 V; 100 W each, 200 W total.
(b) 0.42 A in each; 60 V; 25 W each, 50 W total.
ES   Difficult

19.25: What different resistances can be obtained by using two 2-Ω resistors and one 4-Ω resistor?

Answer: 0.8, 1.5, 2.0, 3.3, 5, 8 Ω
ES   Difficult

19.26: Two 4-Ω resistors are connected in parallel, and this combination is connected in series with 3 Ω. What is the effective resistance of this combination?
  a) 1.2 Ω   b) 5 Ω   c) 7 Ω   d) 11 Ω

Answer: (b)
MC   Moderate

19.27: A 3-Ω resistor is connected in parallel with a 6-Ω resistor. This combination is connected in series with a 4-Ω resistor. The resistors are connected to a 12-volt battery. How much power is dissipated in the 3-Ω resistor?
  a) 2.7 W   b) 5.3 W   c) 6 W   d) 12 W

Answer: (b)
MC   Difficult

19.28: When resistors are connected in parallel, we can be certain that
  a) the same current flows in each one.
  b) the potential difference across each is the same.
  c) the power dissipated in each is the same.
  d) their equivalent resistance is greater than the resistance of any one of the individual resistances.

Answer: (b)
MC   Easy

19.29: When resistors are connected in series,
   a) the same power is dissipated in each one.
   b) the potential difference across each is the same.
   c) the current flowing in each is the same.
   d) more than one of the above is true.

Answer: (c)
   MC   Easy

19.30: Consider three identical resistors, each of resistance R. The maximum power each can dissipate is P. Two of the resistors are connected in series, and a third is connected in parallel with these two. What is the maximum power this network can dissipate?
   a) 2P/3   b) 3P/2   c) 2P   d) 3P

Answer: (b)
   MC   Difficult

19.31: What is the maximum number of different resistance values obtainable by using three resistors, if all three need not be used?
   a) 6   b) 9   c) 11   d) 17

Answer: (d)
   MC   Difficult

19.32: A 3-Ω resistor is connected in parallel with a 6-Ω resistor. This pair is then connected in series with a 4-Ω resistor. These resistors are connected to a battery. What will happen if the 3-Ω resistor breaks, i.e., becomes an infinite resistance?
   a) The current in the 4-Ω resistor will drop to zero.
   b) The current in the 6-Ω resistor will increase.
   c) The current provided by the battery will not change.
   d) The power dissipated in the circuit will increase.

Answer: (b)
   MC   Difficult

19.33:

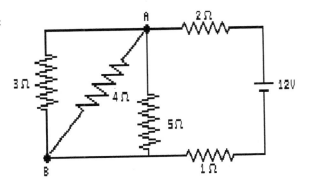

For the circuit shown, determine: (a) the current in each resistor (b) the potential difference between points A and B.

Answer: (a) 2.8 A in 1-Ω and 2-Ω resistors,
1.19 A in 3-Ω resistor,
0.90 A in 4-Ω resistor,
0.72 A in 5-Ω resistor.
(b) 3.58 V between A and B
ES Difficult

19.34:

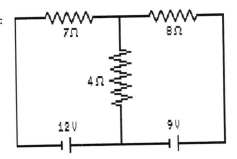

Determine the current and its direction, in each resistor, for the circuit shown here.

Answer: 1.09 A to left in 7 Ω,
0.75 A to right in 8 Ω,
1.84 A up in 4 Ω.
ES Moderate

19.35:

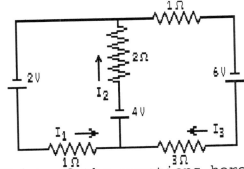

Which of the equations here is valid for the circuit shown?
a) $2 - I_1 - 2I_2 = 0$
c) $4 - I_1 + 4I_3 = 0$
e) $6 - I_1 - 2I_2 = 0$
b) $2 - 2I_1 - 2I_2 - 4I_3 = 0$
d) $-2 - I_1 - 2I_2 = 0$

Answer: (d)
MC Moderate

19.36: Kirchhoff's junction rule is an example of
a) conservation of energy.   b) conservation of charge.
c) conservation of momentum.   d) none of the above.

Answer: (b)
MC   Easy

19.37: Kirchhoff's voltage rule for a closed loop is an example of
a) conservation of energy.   b) conservation of charge.
c) conservation of momentum.   d) none of the above.

Answer: (a)
MC   Easy

19.38:

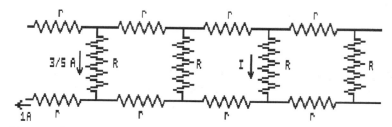

Shown here are a few segments of a circuit used to model an
infinitely long transmission line. The circuit continues
indefinitely in both directions. What is the value, in amperes,
of the current I?
a) 8/125   b) 12/125   c) 16/75   d) 4/25   e) 2/25

Answer: (b)
MC   Difficult

19.39: A 2-μF capacitor is charged to 12 V and then discharged through 4
x 10^6 Ω. How long will it take for the voltage across the
capacitor to drop to 3 V?

Answer: 11.1 s
ES   Moderate

19.40:

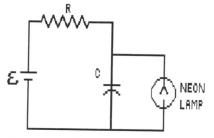

A 4-μF capacitor in series with a
25,000-Ω resistor is charged by a
100-V battery. A neon lamp is
connected across the capacitor (in
parallel with it). A neon lamp
has a very high resistance before
it "fires," i.e. ionizes. When
the voltage across it reaches 70
V, it fires and its resistance
drops almost instantaneously to zero. It then ceases to conduct
once the voltage drops below 70 V. This results in periodically
discharging the capacitor. This is the kind of circuit used to
make roadside warning lights at construction sites. Determine the
frequency at which the light will blink in the above circuit.

Answer: 8.3 Hz
ES   Difficult

19.41: What is the unit for the quantity RC?
   a) Ohms   b) Volt-Ampere/ohm
   c) Seconds   d) Meters

Answer: (c)
        MC   Easy

19.42: A 2-μF capacitor is charged though a 50,000-Ω resistor. How long
       does it take for the capacitor to reach 90% of full charge?
       a) 0.9 s   b) 0.23 s   c) 2.19 s   d) 2.3 s

Answer: (b)
        MC   Moderate

19.43: A 4-μF capacitor is charged to 6 V. It is then connected in
       series with a 3-MΩ resistor and connected to a 12-V battery. How
       long after being connected to the battery will the voltage across
       the capacitor be 9 V?
       a) 5.5 s   b) 8.3 s   c) 12 s   d) 16.6 s

Answer: (b)
        MC   Difficult

19.44: A 4-MΩ resistor is connected in series with a 0.5-μF capacitor.
       The capacitor is initially uncharged. The RC combination is
       charged by a 9-V battery. What is the change in voltage between
       t = RC and t = 3RC?
       a) 11.40 V   b) 7.59 V   c) 5.70 V   d) 2.86 V

Answer: (d)
        MC   Moderate

19.45: A galvanometer has a coil with a resistance of 24 Ω. A current
       of 180 μA causes full-scale deflection. If the galvanometer is
       to be used to construct an ammeter that deflects full scale for
       10 A, what shunt resistor is required?

Answer: 432 $\mu\Omega$
        ES   Moderate

19.46: A galvanometer with a coil resistance of 40 Ω deflects full scale
       for a current of 2 mA. What series resistance should be used
       with this galvanometer in order to construct a voltmeter that
       deflects full scale for 50 V?

Answer: 24960 Ω
        ES   Moderate

19.47: An ideal ammeter should
   a) have a high coil resistance.
   b) introduce a very small series resistance into the circuit whose current is to be measured.
   c) introduce a very large series resistance into the circuit whose current is to be measured.
   d) consist of a galvanometer in series with a large resistor.

Answer: (b)
   MC   Easy

19.48: In order to construct a voltmeter from a galvanometer, one normally would
   a) use a very small shunt resistor.
   b) use a very large shunt resistor.
   c) use a very small series resistor.
   d) use a very large series resistor.

Answer: (d)
   MC   Easy

19.49: A galvanometer with a coil resistance of 80 Ω deflects full scale for a current of 2 mA. What series resistance is required to convert it to a voltmeter reading full scale for 200 V?
   a) 0.0008 Ω b) 99,920 Ω c) 100,000 Ω d) 100,080 Ω

Answer: (b)
   MC   Moderate

19.50: An unknown resistor is wired in series with an ammeter, and a voltmeter is placed in parallel across both the resistor and the ammeter. This network is then placed across a battery. If one computes the value of the resistance by dividing the voltmeter reading by the ammeter reading, the value obtained
   a) is less than the true resistance.
   b) is greater than the true resistance.
   c) is the true resistance.
   d) could be any of the above. It depends on other factors.

Answer: (b)
   MC   Difficult

19.51: An unknown resistor is wired in series with an ammeter, and a voltmeter is placed in parallel across the resistor only. This network is then connected to a battery. If one computes the value of the resistance by dividing the voltmeter reading by the ammeter reading, the value obtained
   a) is less than the true resistance.
   b) is greater than the true resistance.
   c) is the true resistance.
   d) could be any of the above. It depends on other factors.

Answer: (a)
   MC   Difficult

19.52: Increasing the resistance of a voltmeter's series resistance
   a) allows it to measure a larger voltage at full-scale deflection.
   b) allows it to measure a smaller voltage at full-scale deflection.
   c) enables more current to pass through the meter movement at full-scale deflection.
   d) converts it to an ammeter.

Answer: (a)
   MC   Moderate

19.53: A current reading is obtained by properly placing an ammeter in a circuit consisting of one resistor and one battery. As a result,
   a) the voltage drop across the resistor increases.
   b) the current flowing in the circuit increases.
   c) the current flowing in the circuit decreases.
   d) the current flowing in the circuit does not change.

Answer: (c)
   MC   Moderate

19.54: A voltage reading is obtained by placing a voltmeter across a resistor. What happens to the total current flowing in the circuit as a result of this action?
   a) The current increases.
   b) The current decreases.
   c) The current does not change.
   d) The current increases if the meter's internal resistance is less than the original resistance in the circuit and decreases if its internal resistance is greater than the circuit's original resistance.

Answer: (a)
   MC   Moderate

19.55: One amp of current is enough to kill a person.

Answer: True
   TF   Moderate

19.56: A resistor and a capacitor are connected in series to an ideal battery of constant terminal voltage. At the moment contact is made with the battery, the voltage across the resistor is
   a) greater than the battery's terminal voltage.
   b) less than the battery's terminal voltage, but greater than zero.
   c) equal to the battery's terminal voltage.
   d) zero.

Answer: (c)
   MC   Easy

19.57: A resistor and a capacitor are connected in series to an ideal battery of constant terminal voltage. At the moment contact is made with the battery, the voltage across the capacitor is
   a) greater than the battery's terminal voltage.
   b) less than the battery's terminal voltage, but greater than zero.
   c) equal to the battery's terminal voltage.
   d) zero.

Answer: (d)
   MC   Easy

19.58: A resistor and a capacitor are connected in series to an ideal battery of constant terminal voltage. When this system reaches its steady-state, the voltage across the resistor is
   a) greater than the battery's terminal voltage.
   b) less than the battery's terminal voltage, but greater than zero.
   c) equal to the battery's terminal voltage.
   d) zero.

Answer: (d)
   MC   Moderate

19.59: A battery has an internal resistance of 2 Ω. This battery delivers maximum power to a load resistor that has a value of
   a) zero   b) 1 Ω c) 2 Ω d) 4 Ω

Answer: (c)
   MC   Moderate

19.60: Capacitances of 10 μF and 20 μF are connected in parallel, and this pair is then connected in series with a 30 μF capacitor. What is the equivalent capacitance of this arrangement?

Answer: 15 μF
   ES   Moderate

19.61: Consider capacitors C_1, C_2, and C_3, which are connected in series in a closed loop. A switch is placed between C_1 and C_2. With the switch open, C_1 is charged to 12 volts by a battery. The battery is then disconnected and the switch is closed. Determine the final charge on each capacitor, and the potential difference across each given that $C_1 = 2$ μF, and $C_2 = C_3 = 3$ μF.

Answer: $Q_1 = 13.7$ μF; $V_1 = 6.86$ V
        $Q_2 = Q_3 = 10.3$ μC; $V_2 = V_3 = 3.43$ V
        ES   Difficult

19.62: When two or more capacitors are connected in series to a battery,
   a) the total voltage across the combination is the algebraic sum of the voltages across the individual capacitors.
   b) each capacitor carries the same amount of charge.
   c) the equivalent capacitance of the combination is less than the capacitance of any of the capacitors.
   d) All of the above.

Answer: (d)
   MC   Easy

19.63: When two or more capacitors are connected in parallel to a battery,
   a) the voltage across each capacitor is the same.
   b) each capacitor carries the same amount of charge.
   c) the equivalent capacitance of the combination is less than the capacitance of any one of the capacitors.
   d) All of the above.

Answer: (a)
   MC   Easy

19.64: Three identical capacitors are connected in series to a battery. If a total charge of Q flows from the battery, how much charge does each capacitor carry?
   a) 3Q
   c) Q/3
   b) Q
   d) None of the above.

Answer: (b)
   MC   Easy

19.65: Three identical capacitors are connected in parallel to a battery. If a total charge of Q flows from the battery, how much charge does each capacitor carry?
   a) 3Q
   c) Q/3
   b) Q
   d) None of the above.

Answer: (c)
   MC   Easy

19.66: As more and more capacitors are connected in series, the equivalent capacitance of the combination increases.
   a) Always true.
   b) Sometimes true. It depends on the voltage of the battery to which the combination is connected.
   c) Sometimes true. It goes up only if the next capacitor is larger than the average of the existing combination.
   d) Never true.

Answer: (d)
   MC   Moderate

19.67: As more and more capacitors are connected in parallel, the equivalent capacitance of the combination increases.
   a) Always true.
   b) Sometimes true. It depends on the voltage of the battery to which the combination is connected.
   c) Sometimes true. It goes up only if the next capacitor is larger than the average of the existing combination.
   d) Never true.

Answer: (a)
   MC   Moderate

19.68: Four 16 μF capacitors are connected in series. The equivalant capacitance of this combination is
   a) 64 μF   b) 16 μF
   c) 4 μF   d) None of the above.

Answer: (c)
   MC   Easy

19.69: Four 16 μF capacitors are connected in parallel. The equivalent capacitance of this combination is
   a) 64 μF   b) 16 μF
   c) 4 μF   d) None of the above.

Answer: (a)
   MC   Easy

19.70: A 1 μF and a 2 μF capacitor are connected in series across a 3 V battery. What is the voltage across the 2 μF capacitor?
   a) 3 V   b) 2 V
   c) 1 V   d) None of the above.

Answer: (c)
   MC   Moderate

19.71: A 1 μF and a 2 μF capacitor are connected in series across a 3 V battery. What is the voltage across the 1 μF capacitor?
   a) 3 V   b) 2 V
   c) 1 V   d) None of the above.

Answer: (b)
   MC   Moderate

19.72: 5 μF, 10 μF, and 50 μF capacitors are connected in series across a 12 V battery. How much charge is stored in the 5μF capacitor?

Answer: 37.5 μC
   ES   Moderate

19.73: 5 μF, 10 μF, and 50 μF capacitors are connected in series across a 12 V battery. What is the potential difference across the 10 μF capacitor?

Answer: 3.75 V
ES Moderate

19.74: A 120 V battery is used to separately charge both a 1 μF and a 2 μF capacitor. The battery is removed. The fully charged capacitors are connected to each other such that terminals of opposite sign are joined together. Afterwards, what is the equilibrium voltage across one of the capacitors?

Answer: 40 V
ES Difficult

19.75: 1 μF, 2 μF, and 3 μF capacitors are connected in parallel across a 24 V battery. How much energy is stored in this combination when the capacitors are fully charged?

Answer: 1.73 mJ
ES Moderate

19.76: A 2 Ω resistor is in series with a parallel combination of 4 Ω, 6 Ω, and 12 Ω. What is the equivalent resistance of this combination?
   a) 24 Ω b) 4 Ω
   c) 1.83 Ω d) None of the above.

Answer: (b)
MC Easy

19.77: A 14 A current flows into a parallel combination of a 3 Ω and a 4 Ω resistor. What current flows through the 3 Ω resistor?
   a) 8 A   b) 7 A
   c) 6 A   d) None of the above.

Answer: (a)
MC Moderate

19.78: A 14 A current flows into a parallel combination of a 3 Ω and a 4 Ω resistor. What current flows through the 4 Ω resistor?
   a) 8 A   b) 7 A
   c) 6 A   d) None of the above.

Answer: (c)
MC Moderate

19.79: A 22 A current flows into a parallel combination of a 4 Ω, 6 Ω, and 12 Ω resistor. What current flows through the 4 Ω resistor?
  a) 18 A   b) 7.33 A
  c) 3.66 A   d) None of the above.

Answer: (d)
  MC   Moderate

19.80: A 22 A current flows into a parallel combination of a 4 Ω, 6 Ω, and 12 Ω resistors. What current flows through the 6 Ω resistor?
  a) 18 A   b) 7.33 A
  c) 3.66 A   d) None of the above.

Answer: (b)
  MC   Moderate

19.81: A 22 A current flows into a parallel combination of 4 Ω, 6 Ω, and 12 Ω resistors. What current flows through the 12 Ω resistor?
  a) 18 A   b) 7.33 A
  c) 3.66 A   d) None of the above.

Answer: (c)
  MC   Moderate

19.82: Which of the following is a correct statement of Kirchhoff's loop rule?
  a) The net current flowing into a junction must vanish.
  b) The algebraic sum of the emf's around any closed loop must equal zero.
  c) The algebraic sum of the potential drops around any closed loop must equsl zero.
  d) The algebraic sum of the emf's must equal the algebraic sum of the potential drops around any loop.

Answer: (d)
  MC   Easy

19.2: **TABLE 19-2**

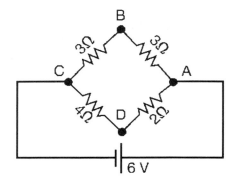

19.83: Refer to Table 19-2.
What is the potential of point A relative to point C?
a) +6 V   b) +4 V
c) +3 V   d) None of the above.

Answer: (a)
MC   Easy   Table: 2

19.84: Refer to Table 19-2.
What is the potential of point B relative to point C?
a) +6 V   b) +4 V
c) +3 V   d) None of the above.

Answer: (c)
MC   Moderate   Table: 2

19.85: Refer to Table 19-2.
What is the potential of point D relative to point C?
a) +6 V   b) +4 V
c) +3 V   d) None of the above.

Answer: (b)
MC   Moderate   Table: 2

19.86: Refer to Table 19-2.
What is the potential of point D relative to point B?
a) +6 V   b) +4 V
c) +3 V   d) None of the above.

Answer: (d)
MC   Moderate   Table: 2

19.3: **TABLE 19-3**

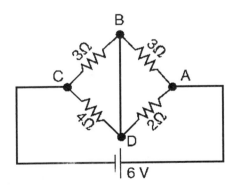

19.87: Refer to Table 19-3.
What current flows from point D to point B?

Answer: 0.353 A
ES   Difficult   Table: 3

19.88: Refer to Table 19-3.
What current flows from the battery?

Answer: 2.06 A
ES Difficult Table: 3

19.89: Refer to Table 19-3.
What is the potential drop from point A to point B?

Answer: 2.47 V
ES Difficult Table: 3

19.90: A galvanometer can be converted to an ammeter by the addition of
a
a) small resistance in parallel.
b) large resistance in parallel.
c) small resistance in series.
d) large resistance in series.

Answer: (a)
MC Easy

19.91: Decreasing the resistance of an ammeter's shunt resistance
a) allows it to measure a larger current at full scale
deflection.
b) allows it to measure a smaller current at full scale
deflection.
c) enables more current to pass directly through the
galvanometer.
d) converts it to a voltmeter.

Answer: (a)
MC Moderate

19.92: A galvanometer has an internal resistance of 100 Ω and deflects
full-scale at 2 mA. What size resistor should be added to it to
convert it to a milliameter capable of reading up to 4 mA?
a) 50 Ω in series   b) 50 Ω in parallel
c) 100 Ω in series   d) 100 Ω in parallel

Answer: (d)
MC Moderate

19.93: A galvanometer has an internal resistance of 100 Ω and deflects
full-scale at 2 mA. What size resistor should be added to it to
convert it to a millivoltmeter capable of reading up to 400 mV?
a) 50 Ω in series   b) 50 Ω in parallel
c) 100 Ω in series   d) 100 Ω in parallel

Answer: (c)
MC Moderate

19.4: **TABLE 19-4**

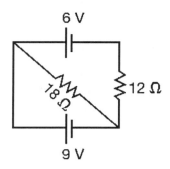

19.94: Refer to Table 19-4.
What current flows in the 12 Ω resistor?

Answer: 0.25 A
ES Difficult Table: 4

19.95: Refer to Table 19-4.
What current flows in the 18 Ω resistor?

Answer: 0.5 A
ES Difficult Table: 4

19.96: Refer to Table 19-4. What current flows in the solid wire
connecting the upper left and lower left corners?

Answer: 0.75 A
ES Difficult Table: 4

19.5: **TABLE 19-5**

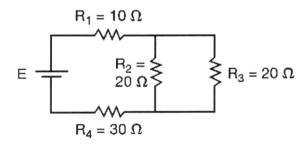

19.97: Refer to Table 19-5. What is the total circuit resistance?
a) 80 Ω   b) 55 Ω   c) 50 Ω   d) 35 Ω

Answer: (b)
MC Moderate Table: 5

19.98: Refer to Table 19-5. If E = 40 V, what is the voltage on R_1?
     a) 6.7 V   b) 8 V   c) 10 V   d) 20 V

Answer: (b)
     MC Moderate Table: 5

19.99: Refer to Table 19-5. If E = 20 V, what is the current through R_3?
     a) 0.05 A   b) 0.2 A   c) 1 A   d) 4 A

Answer: (b)
     MC Moderate Table: 5

19.100: Refer to Table 19-5. If 1.5 A flows through R_2, what is E?
     a) 150 V   b) 75 V   c) 60 V   d) 30 V

Answer: (a)
     MC Difficult Table: 5

19.6: **TABLE 19-6**

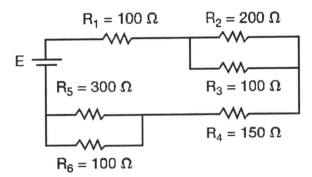

19.101: Refer to Table 19-6. What is the total circuit resistance?
     a) 950 Ω   b) 450 Ω   c) 392 Ω   d) 257 Ω

Answer: (c)
     MC Moderate Table: 6

19.102: Refer to Table 19-6. If E = 100 V, what is the voltage across R_5?
     a) 19.1 V   b) 40 V   c) 75 V   d) 76.6 V

Answer: (a)
     MC Difficult Table: 6

19.103: Refer to Table 19-6. If E = 4 V, what is the current through R_6?
     a) 0.0077 A   b) 0.0173 A   c) 0.04 A   d) 4.0 A

Answer: (a)
     MC Difficult Table: 6

Magnetism

20.1: MATCH THE UNIT TO THE PHYSICAL QUANTITY.

20.1: permeability

Answer: T-m/A
      MA   Easy

20.2: magnetic dipole moment

Answer: A-m^2
      MA   Easy

20.3: magnetic field

Answer: Tesla
      MA   Easy

20.4: turns per meter

Answer: m^{-1}
      MA   Easy

20.5: torque

Answer: N-m^2
      MA   Easy

20.6: Breaking a bar magnet in half will produce a north magnetic monopole and a south magnetic monopole.

Answer: False
      TF Easy

20.7: A proton moving at 5×10^4 m/s horizontally enters a region where a magnetic field of 0.12 T is present, directed vertically downward. What force acts on the proton?

Answer: 9.6×10^{-16} N
      ES Easy

20.8: An electron traveling due north with speed 4×10^5 m/s enters a region where the earth's magnetic field has the magnitude 5×10^{-5} T and is directed downward at 45° below horizontal. What force acts on the electron?

Answer: 2.26×10^{-18} N
      ES Moderate

20.9: Two long parallel wires carry currents of 10 A in opposite directions. They are separated by 40 cm. What is the magnetic field in the plane of the wires at a point that is 20 cm from one wire and 60 cm from the other?

Answer: 6.67×10^{-6} N
ES Moderate

20.10: What is the magnetic field at the center of a circular loop of wire of radius 4 cm when a current of 2 A flows in the wire?

Answer: 3.14×10^{-5} N
ES Moderate

20.11: A very long straight wire carrying 2 A lies along the x-axis. A second very long parallel wire carrying 3 A in the same direction is positioned parallel to the x-axis and passes through the point (0, 1m, 0). What is the magnitude of the magnetic field at the point (0, 0, 1m)?

Answer: 7.62×10^{-7} N
ES Difficult

20.12: A proton is accelerated from rest through 500 V. It enters a magnetic field of 0.30 T oriented perpendicular to its direction of motion. Determine the radius of the path it follows.

Answer: 1.1 cm
ES Moderate

20.13: The SI unit of magnetic field is the
    a) weber.   b) gauss.   c) tesla.   d) lorentz.

Answer: (c)
MC Easy

20.14: A vertical wire carries a current straight up in a region where the magnetic field vector points due north. What is the direction of the resulting force on this current?
    a) Down   b) North   c) East   d) West

Answer: (d)
MC Easy

20.15: A charged particle is injected into a uniform magnetic field such that its velocity vector is perpendicular to the magnetic field vector. Ignoring the particle's weight, the particle will
a) move in a straight line.
b) follow a spiral path.
c) move along a parabolic path.
d) follow a circular path.

Answer: (d)
MC   Easy

20.16: A charged particle is observed traveling in a circular path in a uniform magnetic field. If the particle had been traveling twice as fast, the radius of the circular path would be
a) twice the original radius.
b) four times the original radius.
c) one-half the original radius.
d) one-fourth the original radius.

Answer: (a)
MC   Easy

20.17: A particle carrying a charge of +1 e travels in a circular path in a uniform magnetic field. If instead the particle carried a charge of +2 e, the radius of the circular path would have been
a) twice the original radius.
b) four times the original radius.
c) one-half the original radius.
d) one-fourth the original radius.

Answer: (c)
MC   Easy

20.18: At double the distance from a long current-carrying wire, the strength of the magnetic field produced by that wire decreases to
a) 1/8 of its original value.   b) 1/4 of its original value.
c) 1/2 of its original value.   d) none of the above.

Answer: (c)
MC   Easy

20.19: A horizontal wire carries a current straight toward you. From your point of view, the magnetic field caused by this current
a) points directly away from you.
b) points to the left.
c) circles the wire in a clockwise direction.
d) circles the wire in a counter-clockwise direction.

Answer: (d)
MC   Easy

20.20: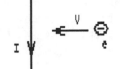

A wire lying in the plane of the page carries a current toward the bottom of the page, as shown. What is the direction of the magnetic force it produces on an electron that is moving perpendicularly toward the wire, also in the plane of the page, from your right?
   a) Zero
   b) Perpendicular to the page and towards you
   c) Perpendicular to the page and away from you
   d) Toward the top of the page
   e) Toward the bottom of the page

Answer: (e)
   MC   Moderate

20.21: Consider two current-carrying circular loops. Both are made from one strand of wire and both carry the same current, but one has twice the radius of the other. Compared to the magnetic field at the center of the smaller loop, the magnetic field at the center of the larger loop is
   a) 8 times stronger.   b) 4 times stronger.
   c) 2 times stronger.   d) none of the above.

Answer: (d)
   MC   Moderate

20.22: Parallel wires carrying currents in the same direction will repel each other.

Answer: False
   TF   Moderate

20.23: What is the force per meter on a straight wire carrying 5 A when it is placed in a magnetic field of 0.02 T? The wire makes an angle of 27° with respect to the magnetic field lines.
   a) 0.022 N   b) 0.045 N   c) 0.17 N   d) 0.26 N

Answer: (b)
   MC   Easy

20.24: A solenoid 20 cm long is wound with 5000 turns of wire. What magnetic field is produced at the center of the solenoid when a current of 10 A flows?
   a) 1.6 T   b) 0.84 T   c) 0.67 T   d) 0.31 T

Answer: (d)
   MC   Moderate

20.25: When a ferromagnetic material is placed in an external magnetic field, the net magnetic field of its magnetic domain becomes smaller.

Answer: False
       TF   Easy

20.26: What is the magnetic moment of a rectangular loop of 120 turns that carries 6 A if its dimensions are 4 cm x 8 cm?

Answer: 2.30 A-m^2
       ES   Easy

20.27: Four long parallel wires each carry 2 A in the same direction. They are parallel to the z-axis, and they pass through the corners of a square of side 4 cm positioned in the x-y plane. What magnetic field does one of the wires experience due to the other wires?

Answer: 2.12 x 10^{-5} T
       ES Difficult

20.28: A circular loop of wire of cross-sectional area 0.12 m^2 consists of 200 turns, each carrying 0.5 A. It is placed in a magnetic field of 0.05 T oriented at 30° to the plane of the loop. What torque acts on the loop?

Answer: 0.52 N-m
       ES Moderate

20.29: A charged particle moves with a constant speed through a region where a uniform magnetic field is present. If the magnetic field points straight upward, the magnetic force acting on this particle will be maximum when the particle moves
   a) straight upward.
   b) straight downward.
   c) upward at an angle of 45° above the horizontal.
   d) none of the above.

Answer: (d)
       MC Moderate

20.30: At a particular instant, a proton moves eastward at speed V in a uniform magnetic field that is directed straight downward. The magnetic force that acts on it is
   a) zero.   b) directed upward.
   c) to the south.   d) none of the above.

Answer: (d)
       MC Easy

20.31: At a particular instant, an electron moves eastward at speed V in a uniform magnetic field that is directed straight downward. The magnetic force that acts on it is
a) zero.
b) directed upward.
c) directed to the south.
d) none of the above.

Answer: (c)
MC   Moderate

20.32: A charged particle moves and experiences no magnetic force. From this we can conclude that
a) no magnetic field exists in that region of space.
b) the particle is moving parallel to the magnetic field.
c) the particle is moving at right angles to the magnetic field.
d) either no magnetic field exists or the particle is moving parallel to the field.

Answer: (d)
MC   Moderate

20.33: Consider two current-carrying circular loops. Both are made from one strand of wire and both carry the same current, but one has twice the radius of the other. Compared to the magnetic moment of the smaller loop, the magnetic moment of the larger loop is
a) 16 times stronger.
b) 8 times stronger.
c) 4 times stronger.
d) 2 times stronger.

Answer: (c)
MC   Moderate

20.34: A charged particle moving along a direction perpendicular to the plane of a loop of current-carrying wire, and which passes through the loop's center, will experience no magnetic force.

Answer: True
TF   Moderate

20.35: What fundamental fact underlies the operation of essentially all electric motors?
a) Opposite electric charges attract and like charges repel.
b) A current-carrying conductor placed perpendicular to a magnetic field will experience a force.
c) Alternating current and direct current are both capable of doing work.
d) Iron is the only element that is magnetic.
e) A magnetic north pole carries a positive electric charge, and a magnetic south pole carries a negative electric charge.

Answer: (b)
MC   Easy

20.36: A motor is a device that converts mechanical energy into electrical energy.

Answer: False
TF   Easy

20.37: A split-ring commutator in a DC motor reverses the polarity and current each half-cycle.

Answer: True
TF   Easy

20.38: The torque on the rotating loop of wire in a DC motor is constant, in magnitude, as the loop rotates.

Answer: False
TF   Moderate

20.39: In a mass spectrometer a particle of mass m and charge q is accelerated through a potential difference V and allowed to enter a magnetic field B, where it is deflected in a semi-circular path of radius R. The magnetic field is uniform and oriented perpendicular to the velocity of the particle. Derive an expression for the mass of the particle in terms of B, q, V, and R.

Answer: $m = qB^2 R^2/2V$
ES   Moderate

20.40: A beam of electrons is accelerated through a potential difference of 10 kV before entering a velocity selector. If the B-field of the velocity selector has a value of 10^{-2} T, what value of the E-field is required if the particles are to be undeflected?
a) 2.26×10^3 V/m   b) 5.93×10^5 V/m
c) 6.04×10^5 V/m   d) 7.24×10^6 V/m

Answer: (b)
MC   Moderate

20.41: A doubly charged ion with velocity 6×10^6 m/s moves in a path of radius 30 cm in a magnetic field of 0.8 T in a mass spectrometer. What is the mass of this ion?
a) 12.8×10^{-27} kg   b) 6.68×10^{-27} kg
c) 3.34×10^{-27} kg   d) 8.22×10^{-27} kg

Answer: (a)
MC   Moderate

20.42: 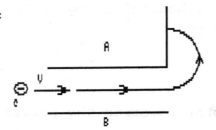 An electron enters the velocity selector of a mass spectrometer in the direction shown in the sketch. If it emerges along the path shown, which plate (A or B) has the higher potential?

Answer: Plate B
ES   Moderate

20.43: The earth's magnetic field at its magnetic equator is approximately 10^4 T.

Answer: False
TF   Easy

20.44: A proton with velocity 2 x 10^5 m/s enters a region where a uniform magnetic field of 0.3 T is present. The proton's velocity makes an angle of 60° with the magnetic field. What is the pitch of the helical path followed by the proton?

(The "pitch" is the distance traveled parallel to the magnetic field while making one complete revolution around a magnetic field line.)
   a) 0.022 m   b) 0.077 m   c) 0.140 m   d) 0.250 m

Answer: (a)
MC   Difficult

20.45: At a particular instant a proton moves eastward at speed v in a uniform magnetic field that is directed straight down. The magnetic force that acts on the proton is
   a) zero.   b) directed upward.
   c) directed to the south.   d) None of the above.

Answer: (d)
MC   Easy

20.46: A vertical wire carries a current straight down. To the east of this wire, the magnetic field points
   a) north.   b) east.   c) south.   d) down.

Answer: (c)
MC   Easy

20.47: Two long parallel wires are placed side-by-side on a horizontal table. If the wires carry current in the same direction,
a) one wire is lifted slightly as the other wire is forced against the table's surface.
b) both wires are lifted slightly.
c) the wires attract each other.
d) the wires repel each other.

Answer: (c)
MC   Easy

20.48: Two long parallel wires are placed side-by-side on a horizontal table. If the wires carry current in opposite directions,
a) one wire is lifted slightly as the other is forced against the table's surface.
b) both wires are lifted slightly.
c) the wires attract each other.
d) the wires repel each other.

Answer: (d)
MC   Easy

20.49: The maximum torque on a current carrying loop occurs when the angle between the loop's magnetic moment and the magnetic field vector is
a) 0°   b) 90°
c) 180°   d) None of the above.

Answer: (b)
MC   Moderate

20.50: A current carrying circular loop of wire lies flat on a table top. When viewed from above, the current moves around the loop in a counterclockwise sense. What is the direction of the magnetic field caused by this current, inside the loop? The magnetic field
a) circles the loop in a clockwise direction.
b) circles the loop in a counterclockwise direction.
c) points straight up.
d) points straight down.

Answer: (c)
MC   Easy

20.51: A current carrying loop of wire lies flat on a table top. When viewed from above, the current moves around the loop in a counterclockwise sense. What is the direction of the magnetic field caused by this current, outside the loop? The magnetic field
a) circles the loop in a clockwise direction.
b) circles the loop in a counterclockwise direction.
c) points straight up.
d) points straight down.

Answer: (d)
MC   Easy

20.52: A current carrying circular loop of wire lies flat on a table top. When viewed from above, the current moves around the loop in a counterclockwise sense. What is the direction of the loop's magnetic moment? The magnetic moment
a) circles the loop in a clockwise direction.
b) circles the loop in a counterclockwise direction.
c) points straight up.
d) points straight down.

Answer: (c)
MC   Easy

20.53: Two long parallel wires placed side-by-side on a horizontal table carry identical current straight toward you. From your point of view, the magnetic field at the point exactly between the two wires
a) points up.   b) points down.
c) points toward you.   d) is zero.

Answer: (d)
MC   Moderate

20.54: Two long parallel wires placed side-by-side on a horizontal table carry identical size currents in opposite directions. The wire on your right carries current toward you, and the wire on your left carries current away from you. From your point of view, the magnetic field at the point exactly midway between the two wires
a) points up.   b) points down.
c) points toward you.   d) is zero.

Answer: (b)
MC   Moderate

20.55: When placed askew in a magnetic field, a current carrying loop that is free to rotate in any direction will experience a torque until its magnetic moment vector
a) is at right angles to the magnetic field vector.
b) makes a 45° angle with the magnetic field vector.
c) makes an angle of 270° with the magnetic field vector.
d) is aligned with the magnetic field vector.

Answer: (d)
MC   Moderate

20.56: A charged particle moves across a constant magnetic field. The magnetic force on this particle
a) changes the particle's speed.
b) causes the particle to accelerate.
c) is in the direction of the particle's motion.
d) Both (a) and (b) are correct.

Answer: (b)
MC   Moderate

20.57: An electron has an initial velocity to the south but is observed to curve upward as the result of a magnetic field. The direction of the magnetic field is
a) to the west.   b) to the east.
c) upward.   d) downward.

Answer: (a)
MC   Moderate

20.58: A proton travels at a speed of 5×10^7 m/s through a 1.0 T magnetic field. What is the magnitude of the magnetic force which acts on the proton if the angle between the proton's velocity and the magnetic field vector is 30°?

Answer: 4.0×10^{-12} N
ES   Easy

20.59: A very long straight wire carries a current of 25 A. What is the magnitude of the magnetic field at a distance of 0.15 m from the wire?

Answer: 3.33×10^{-5} T
ES   Easy

20.60: A circular wire loop of area $0.25 \, m^2$ carries a current of 5 A. The coil lies in a horizontal plane with the current flowing in the counterclockwise direction when viewed from above. At this point, the earth's magnetic field is 1.2×10^{-4} T directed 60° below the horizontal. What is the magnitude of the torque which acts on the loop?

Answer: 7.5×10^{-5} N·m
ES   Moderate

20.61: A circular wire loop lies in a horizontal plane on a table and carries current in a counterclockwise direction when viewed from above. At this point, the earth's magnetic field points to the north and dips below the horizontal. Which side of the coil tends to lift off of the table because of the torque caused by the current's interaction with the magnetic field?
a) The north side.
b) The east side.
c) The south side.
d) The west side.

Answer: (c)
MC   Moderate

20.62: The north pole of a magnet points toward the earth's
a) north pole.
b) south pole.
c) center.
d) middle latitudes.

Answer: (a)
MC   Easy

20.63: The earth's magnetic north pole is magnetically
a) north.
b) south.
c) neutral.

Answer: (b)
MC   Easy

20.64: Which of the following is correct?
a) When a current carrying wire is in your right hand, thumb in the direction of the magnetic field lines, your fingers point in the direction of the current.
b) When a current carrying wire is in your left hand, thumb in the direction of the magnetic field lines, your fingers point in the direction of the current.
c) When a current carrying wire is in your right hand, thumb in direction of the current, your fingers point in the direction of the magnetic field lines.
d) When a current carrying wire is in your left hand, thumb pointing in the direction of the current, your fingers point in the direction of the magnetic field lines.

Answer: (c)
MC   Easy

20.65: What is the strength of a magnetic field 4 cm from a long straight wire carrying 3 A of current?
a) 3.7×10^{-6} T
b) 4.8×10^{-6} T
c) 1.5×10^{-5} T
d) 4.6×10^{-5} T

Answer: (c)
MC   Easy

20.66: A high power line carrying 1000 A generates what magnetic field at the ground, 10 m away?
   a) 4.7×10^{-6} T
   b) 6.4×10^{-6} T
   c) 2.0×10^{-5} T
   d) 5.6×10^{-5} T

Answer: (c)
   MC   Easy

20.67: What is the magnetic field at the center of a solenoid that is 20 cm long and has 1000 turns, when 1 A of current flows?
   a) 0.063 T
   b) 0.197 T
   c) 3.2 T
   d) 4.8 T

Answer: (a)
   MC   Easy

20.68: How much current must flow for 1×10^{-3} T of magnetic field to be present 1 cm from a wire?
   a) 0.05 A
   b) 9.15 A
   c) 15.9 A
   d) 50.0 A

Answer: (d)
   MC   Moderate

20.69: How much current must pass through a 400 turn coil 4 cm long to generate a 1.0 T magnetic field at the center?
   a) 0.013 A
   b) 13.0 A
   c) 39.5 A
   d) 79.6 A

Answer: (d)
   MC   Moderate

20.70: How many turns should a 10 cm long solenoid have if it is to generate a 1.5×10^{-3} T magnetic field on 1 A of current?
   a) 12
   b) 15
   c) 119
   d) 1194

Answer: (c)
   MC   Moderate

20.71: 1 T is equivalent to
   a) 1 N·m/A
   b) 1 N·A/m
   c) V·m/A
   d) N/A·m

Answer: (d)
   MC   Moderate

20.72: A velocity selector consists of a charged particle passing through crossed electric and magnetic fields. The forces exerted by these fields are in opposite directions, and only particles of a certain velocity will move in a straight line. In the following, diregard the magnitudes of the fields. In a velocity selector, the particles move toward the east, and the magnetic field is directed to the north. What direction should the electric field point?

   a) east   b) west   c) up   d) down

Answer: (d)
   MC   Moderate

20.73: A long straight wire carries current toward the east. A proton moves toward the east alongside and just south of the wire. What is the direction of the force on the proton?

   a) north   b) south   c) up   d) down

Answer: (a)
   MC   Moderate

Electromagnetic Induction and Faraday's Law; AC Circuits

21.1: **MATCH THE UNIT TO THE PHYSICAL QUANTITY.**

21.1: back emf

Answer: volts
       MA   Easy

21.2: magnetic flux

Answer: weber
       MA   Easy

21.3: magnetic field

Answer: tesla
       MA   Easy

21.4: All of the following are units of magnetic flux except:
    a) T-m^2 b) T/volt-m
    c) weber   d) volt-second

Answer: (b)
       MC   Moderate

21.5: Magnetic flux density has the same units as
    a) magnetic flux.   b) back emf.
    c) magnetic dipole.   d) magnetic field.

Answer: (d)
       MC   Moderate

21.6: A coil of 40 turns and cross-sectional area 12 cm^2 is oriented
      perpendicular to a magnetic field, which varies from zero to 1.2
      T in 0.02 s. What emf is induced in the coil?

Answer: 2.88 V
       ES   Easy

21.7: A circular loop of one turn and radius 5 cm is positioned with
      its axis parallel to a magnetic field of 0.6 T. By means of high
      explosives, the area of the loop is suddenly reduced to
      essentially zero in 0.5 ms. What emf is induced in the loop?

Answer: 9.4 V
       ES   Moderate

21.8: An eagle, with a wingspread of 2 m, flies due north at 8 m/s in a region where the vertical component of the earth's magnetic field is 0.2 x 10^{-4} T. What emf would be developed between the eagle's wing tips? (It has been speculated that this phenomenon could play a role in the navigation of birds, but the effect is too small, in all likelihood.)

Answer: 0.32 mV
ES   Easy

21.9: A coil of 160 turns and area 0.20 m^2 is placed with its axis parallel to a magnetic field of 0.4 T. The magnetic field changes from 0.4 T in the x-direction to 0.4 T in the negative x-direction in 2 s. If the resistance of the coil is 16 Ω, at what rate is power generated in the coil?

Answer: 10.2 W
ES   Moderate

21.10: A wire moves across a magnetic field. The emf produced in the wire depends on`
a) the field's magnetic flux.
b) the length of the wire.
c) the orientation of the wire with respect to the magnetic field vector.
d) all of the above.

Answer: (d)
MC   Easy

21.11: Faraday's law of induction states that the emf induced in a loop of wire is proportional to
a) the magnetic flux.
b) the magnetic flux density times the loop's area.
c) the time variation of the magnetic flux.
d) current divided by time.

Answer: (c)
MC   Easy

21.12: Doubling the number of loops of wire in a coil produces what kind of change on the induced emf, assuming all other factors remain constant?
a) The induced emf is 4 times as much.
b) The induced emf is 3 times as much.
c) The induced emf is twice as much.
d) There is no change in the induced emf.

Answer: (c)
MC   Easy

21.13: The flux through a coil changes from 4×10^{-5} Wb to 5×10^{-5} Wb in 0.1 s. What emf is induced in this coil?
    a) 5×10^{-4} V   b) 4×10^{-4} V
    c) 1×10^{-4} V   d) None of the above

Answer: (c)
    MC   Easy

21.14: A flux of 4×10^{-5} Wb is maintained through a coil for 0.5 s. What emf is induced in this coil by this flux?
    a) 8×10^{-5} V
    b) 4×10^{-5} V
    c) 2×10^{-5} V
    d) No emf is induced in this coil.

Answer: (d)
    MC   Moderate

21.15: A coil lies flat on a table top in a region where the magnetic field vector points straight up. The magnetic field vanishes suddenly. When viewed from above, what is the sense of the induced current in this coil as the field fades?
    a) The induced current flows counterclockwise.
    b) The induced current flows clockwise.
    c) There is no induced current in this coil.
    d) The current flows clockwise initially, and then it flows counterclockwise before stopping.

Answer: (a)
    MC   Moderate

21.16: A circular loop of wire is rotated at constant angular speed about an axis whose direction can be varied. In a region where a uniform magnetic field points straight down, what must be the orientation of the loop's axis of rotation if the induced emf isto be zero?
    a) Any horizontal orientation will do.
    b) It must make an angle of 45° to the vertical.
    c) It must be vertical.
    d) None of the above.

Answer: (c)
    MC   Moderate

21.17: A horizontal metal bar rotates at a constant angular velocity ω about a vertical axis through one of its ends while in a constant magnetic field B that is directed down. The emf induced between the two ends of the bar is
    a) constant and proportional to the product $B\,\omega$.
    b) constant and proportional to the product $B\,\omega^2$.
    c) constant and proportional to the product $B^2\,\omega^2$.
    d) none of the above.

Answer: (a)
    MC   Moderate

21.18: A horizontal metal bar that is 2 m long rotates at a constant angular velocity of 2 rad/s about a vertical axis through one of its ends while in a constant magnetic field of 5×10^{-5} T. If the magnetic field vector points straight down, what emf is induced between the two ends of the bar?

a) 5.6×10^{-3} V
b) 3×10^{-3} V
c) 2×10^{-4} V
d) 1.6×10^{-4} V

Answer: (c)
MC   Moderate

21.19: Consider a circular ring of material through which some magnetic field lines pass. In order to induce an emf around the ring, it is necessary that

a) the material be a conductor.
b) the material be an insulator.
c) the number of magnetic field lines passing through the ring be increased.
d) the ring be rotated.
e) None of the above statements is necessarily true.

Answer: (e)
MC   Moderate

21.20: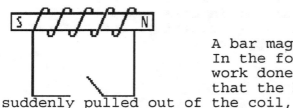
A bar magnet is positioned inside a coil. In the following, "work" refers to any work done as a consequence of the fact that the bar is magnetic. If the bar is suddenly pulled out of the coil,

a) no work will be done, independent of whether or not the switch is closed.
b) more work will be done if the switch is open.
c) more work will be done if the switch is closed.
d) equal (non-zero) amounts of work will be done whether or not the switch is open.
e) whether the work done is positive or negative depends on whether the magnet is pulled out the right or left end of the coil.

Answer: (c)
MC   Moderate

21.21: 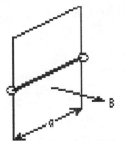 A connecting wire of negligible resistance is bent into the shape shown here. The long portions of the wire are vertical. A bar of length a, of resistance R, and mass m can slide downward without friction on the vertical sections. A uniform horizontal magnetic field B is present. The cross-bar is released from rest, and allowed to fall under the influence of gravity. The vertical wires are long enough so that their length is of no concern. Under these conditions, how will the falling bar behave?
  a) It will continue to accelerate indefinitely.
  b) It will slow down and finally come to rest in equilibrium.
  c) It will reach a constant terminal velocity of $mgR/2a^2B^2$.
  d) It will reach a constant terminal velocity of mgR/a^2B^2.
  e) It will reach a constant terminal velocity of $2mgR/a^2B^2$.

Answer: (d)
  MC Difficult

21.22: A 60-Hz AC generator produces a maximum emf of 170 V. What is the value of the emf 0.025 s after it has its maximum value?

Answer: -170 V
  ES Moderate

21.23: The windings of a DC motor have a resistance of 6 Ω. The motor operates on 120 V AC, and when running at full speed it generates a back emf of 105 V. (a) What is the starting current of the motor? (b) What current does the motor draw when operating at full speed?

Answer: (a) 20 A   (b) 2.5 A
  ES Moderate

21.24: Suppose that you wish to construct a simple AC generator with an output of 12 V maximum when rotated at 60 Hz. A magnetic field of 0.05 T is available. If the area of the rotating coil is 100 cm^2, how many turns are needed?

Answer: 64 turns
  ES Moderate

21.25: A circular loop of wire is rotated at constant angular speed about an axis whose direction can be varied. In a region where a uniform magnetic field points straight down, what must be the orientation of the loop's axis of rotation if the induced emf is to be a maximum?
  a) Any horizontal orientation will do.
  b) It must make an angle of 45° to the vertical.
  c) It must be vertical.
  d) None of the above.

Answer: (a)
  MC Moderate

21.26: A generator coil rotates through 60 revolutions each second. The frequency of the emf is
   a) 30 Hz.
   b) 60 Hz.
   c) 120 Hz.
   d) cannot be determined from the information given.

Answer: (b)
   MC   Easy

21.27: An AC generator has 80 rectangular loops on its armature. Each loop is 12 cm long and 8 cm wide. The armature rotates at 1200 rpm about an axis parallel to the long side. If the loop rotates in a uniform magnetic field of 0.3 T, which is perpendicular to the axis of rotation, what will be the maximum output voltage of this generator?
   a) 19.7 V   b) 26.6 V   c) 29.0 V   d) 35.0 V

Answer: (c)
   MC   Moderate

21.28: The coil of a generator has 50 loops and a cross-sectional area of 0.25 m^2. What is the maximum emf generated by this generator if it is spinning with an angular velocity of 4 rad/s in a 2 T magnetic field?
   a) 100 V   b) 50 V
   c) 400 V   d) None of the above

Answer: (a)
   MC   Easy

21.29: The cross-sectional area of an adjustable single loop is reduced from 1 m^2 to 0.25 m^2 in 0.1 s. What is the average emf that is induced in this coil if it is in a region where B = 2 T upward, and the coil's plane is perpendicular to B?
   a) 12 V   b) 15 V   c) 18 V   d) 21 V

Answer: (b)
   MC   Easy

288

21.30: A lightbulb is plugged into a household outlet, and one of the wires leading to it is wound into a coil. Now a slug of iron is slid into the coil. What effect will this have?
a) The lamp will get brighter because there will be an induced emf that will drive more current through the lamp.
b) The lamp will dim because energy will now be dissipated in the iron.
c) The lamp will dim because a back emf will be produced by the coil, and this will reduce the current flowing in the coil and the lamp.
d) This will have no effect on the brightness of the lamp.
e) This will not affect the lamp's brightness because the coil will shift the phase, but not the magnitude, of the current through the lamp.

Answer: (c)
MC   Moderate

21.31: An ideal transformer has 60 turns on its primary coil and 300 turns on its secondary coil. If 120 V at 2 A is applied to the primary, what voltage and current are present in the secondary?

Answer: 600 V; 0.4 A
ES   Easy

21.32: A step-down transformer is needed to reduce a primary voltage of 120 V AC to 6 V AC. What turns ratio is required?

Answer: 20:1
ES   Easy

21.33: A transformer is a device that
a) operates on either DC or AC.
b) operates only on AC.
c) operates only on DC.
d) increases the power level of a circuit.

Answer: (b)
MC   Easy

21.34: The secondary coil of a neon sign transformer provides 7500 V at 0.01 A. The primary coil operates on 120 V. What does the primary draw?
a) 0.625 A
b) 6.25×10^{-4} A
c) 0.16 A
d) 1.66 A

Answer: (a)
MC   Moderate

21.35: A power transmission line 50 km long has a total resistance of 0.60 Ω. A generator produces 100 V at 70 A. In order to reduce energy loss due to heating of the transmission line, the voltage is stepped up with a transformer with a turns ratio of 100:1. (a) What percentage of the original energy would be lost if the transformer were not used? (b) What percentage of the original energy is lost when the transformer is used?

Answer: (a) 42%   (b) 0.0042%
ES   Difficult

21.36: The output of a generator is 440 V at 20 A. It is to be transmitted on a line with resistance of 0.6 Ω. To what voltage must the generator output be stepped up with a transformer if the power loss in transmission is not to exceed 0.01% of the original power?
   a) 4.4 kV   b) 22 kV   c) 44.5 kV   d) 72.7 kV

Answer: (d)
MC   Difficult

21.37: Alternating current and voltage (AC), rather than direct current and voltage (DC), are used for domestic electrical power. The reason for this is based on one of the following physical principles. Which one?
   a) When a current flows through a resistance, heat is generated.
   b) When a current flows through a wire, a magnetic field is created.
   c) When the number of magnetic field lines passing through a circuit changes, a voltage is generated in the circuit.
   d) When a current-carrying wire is placed in a magnetic field, it may experience a force if properly oriented.
   e) Opposite electric charges attract each other, and like charges repel each other.

Answer: (c)
MC   Moderate

21.38: The coil of a generator rotates through one complete turn in a uniform magnetic field 50 times every second. When this generator is connected to an external load, the current through the load reverses directions
   a) 25 times every second.   b) 50 times every second.
   c) 100 times every second.   d) none of the above.

Answer: (c)
MC   Moderate

21.39: A very long solenoid has 1×10^6 turns per meter and a cross-sectional area of 2×10^{-4} m^2. A short, secondary coil with 1×10^3 turns is wound over the solenoid's midpoint. What emf is induced in the secondary coil if the current in the solenoid is changing at a rate of 4 A/s?
  a) 1.01 V   b) 1.62 V   c) 2.71 V   d) 4.44 V

Answer: (a)
    MC   Moderate

21.40: capacitive reactance

Answer: s/F
    MA   Easy   Table: 1

21.41: power factor

Answer: dimensionless
    MA   Easy   Table: 1

21.42: inductive reactance

Answer: H/s
    MA   Easy   Table: 1

21.43: impedance

Answer: Z (ohms)
    MA   Easy   Table: 1

21.44: inductance

Answer: henrys
    MA   Easy   Table: 1

21.45: capacitance

Answer: farads
    MA   Easy   Table: 1

21.46: All of the following have the same units except:
    a) inductance.   b) capacitive reactance.
    c) impedance.   d) resistance.

Answer: (a)
    MC   Easy

21.47: The polarity of the voltage in a circuit operating at 60 Hz reverses
    a) 30 times per second.   b) 60 times per second.
    c) 90 times per second.   d) 120 times per second.

Answer: (d)
    MC Easy

21.48: A 150-W lamp is placed into a 120-V AC outlet. (a) What are the peak and rms currents? (b) What is the resistance of the lamp?

Answer: (a) 1.77 A; 1.25 A   (b) 96 Ω
    ES Moderate

21.49: The current through a 50-Ω resistor is I = 0.8 sin (240 t). (a) What is the peak current? (b) What is the rms current? (c) At what frequency does the current vary? (d) How much power is dissipated in the resistor?

Answer: (a) 0.8 A   (b) 0.57 A   (c) 38 Hz   (d) 16 W
    ES Moderate

21.50: The peak output of a generator is 20 A and 240 V. What average power is provided by the generator?

Answer: 2400 W
    ES Easy

21.51: The output voltage of a power supply is V = V_O sin (120 t). At t = 0.01 s the output voltage is 4.2 V. What is the rms voltage of the generator?

Answer: 3.19 V
    ES Moderate

21.52: What do the letters "rms" stand for?
    a) Real modulated signal
    b) Rationalized metric system
    c) Reduced mean sin (function)
    d) Root mean square

Answer: (d)
    MC Easy

21.53: The amount of AC current that produces the same average joule heating effect as 1 ampere of DC current is
    a) 1 A peak current.   b) 1 A peak-to-peak current.
    c) $1/\sqrt{2}$ A peak current.   d) $\sqrt{2}$ A peak current.

Answer: (d)
    MC Moderate

21.54: In a purely inductive AC circuit, the current lags behind the voltage by 90°.

Answer: True
TF Moderate

21.55: At what frequency does a 10-μF capacitor have a reactance of 1200 Ω?

Answer: 13.3 Hz
ES Moderate

21.56: 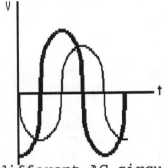 Compare the voltage waves from two different AC circuits. The absolute value of the phase difference is

a) zero   b) 45°   c) 90°   d) 180°

Answer: (c)
MC Easy

21.57: What is the reactance of a 1-mH inductor at 60 Hz?

Answer: 0.377 Ω
ES Easy

21.58: What current flows in a 60-mH inductor when 120 V AC at a frequency of 20 kHz is applied to it?

Answer: 15.9 mA
ES Moderate

21.59: What capacitance will have the same reactance as a 100-mH inductance in a 120-V, 60-Hz circuit?

Answer: 70 μF
ES Moderate

21.60: A resistor is connected to an AC power supply. On this circuit, the current
    a) leads the voltage by 90°.
    b) lags the voltage by 90°.
    c) is in phase with the voltage.
    d) none of the above.

Answer: (c)
    MC   Easy

21.61: A pure inductor is connected to an AC power supply. In this circuit, the current
    a) leads the voltage by 90°.
    b) lags the voltage by 90°.
    c) is in phase with the voltage.
    d) none of the above.

Answer: (b)
    MC   Moderate

21.62: As the frequency of the AC voltage across a capacitor approaches zero, the capacitive reactance of that capacitor
    a) approaches zero.   b) approaches infinity.
    c) approaches unity.   d) none of the above.

Answer: (b)
    MC   Easy

21.63: As the frequency of the AC voltage across an inductor approaches zero, the inductive reactance of that coil
    a) approaches zero.   b) approaches infinity.
    c) approaches unity.   d) none of the above.

Answer: (a)
    MC   Easy

21.64: If the frequency of the AC voltage across a capacitor is doubled, the capacitive reactance of that capacitor
    a) increases to 4 times its original value.
    b) increases to twice its original value.
    c) decreases to one-half its original value.
    d) decreases to one-fourth its original value.

Answer: (c)
    MC   Easy

21.65: If the frequency of the AC voltage across an inductor is doubled, the inductive reactance of that inductor
a) increases to 4 times its original value.
b) increases to twice its original value.
c) decreases to one-half its original value.
d) decreases to one-fourth its original value.

Answer: (b)
MC   Easy

21.66: There is no power loss in a pure LC circuit.

Answer: True
TF   Moderate

21.67: A phase diagram is a graph of phase angle on the vertical axis and resistance on the horizontal axis.

Answer: False
TF   Easy

21.68: Consider an RLC circuit. The impedance of the circuit increases if R increases.
a) Always true.
b) True only if X_L is less than or equal to X_C.
c) True only if X_L is greater than or equal to X_C.
d) Never true.

Answer: (a)
MC   Easy

21.69: Consider an RLC circuit. The impedance of the circuit increases if X_L increases.
a) Always true.
b) True only if X_L is less than or equal to X_C.
c) True only if X_L is greater than or equal to X_C.
d) Never true.

Answer: (c)
MC   Moderate

21.70: A series RC circuit has resistance of 150 Ω and capacitance of 40 μF. It is driven by a 120-V, 60-Hz source. (a) How much power is dissipated in the resistor? (b) How much power is dissipated in the capacitor?

Answer: (a) 88 W   (b) zero
ES   Difficult

21.71: A series RL circuit is driven by a 120-V, 60-Hz source. The values of the resistance and inductance are R = 20 Ω and L = 160 mH. Determine (a) the current. (b) phase angle between the current and the applied voltage.

Answer: (a) I = 1.89 A   (b) I lags V by 72°
ES Difficult

21.72: A 50-Ω resistor is placed in series with a 40-mH inductor. At what frequency will the current in such a circuit lag the applied voltage by exactly 45°?

Answer: 199 Hz
ES Moderate

21.73: The power factor of an RLC circuit is defined as cos ϕ = R/Z, but the phase angle can also be calculated from
a) $\sin \phi = (X_L - X_C)/Z$ b) $\sin \phi = (X_L - X_C)/RZ$
c) $\sin \phi = Z/(X_L - X_C)$ d) $\sin \phi = RZ/(X_L - X_C)$

Answer: (a)
MC Moderate

21.74: A series RLC circuit has R = 20 Ω, L = 200-mH, C = 10 μF. At what frequency should the circuit be driven in order to have maximum power transferred from the driving source?
a) 113 Hz   b) 167 Hz   c) 277 Hz   d) 960 Hz

Answer: (a)
MC Moderate

21.75: What size capacitor must be placed in series with a 30-Ω resistor and a 40-mH coil if the resonant frequency of the circuit is to be 1000 Hz?
a) 0.63 μF   b) 0.50 μF   c) 0.22 μF   d) 0.17 μF

Answer: (a)
MC Moderate

21.76: Consider an RLC circuit that is driven by an AC applied voltage. At resonance,
a) the peak voltage across the capacitor is greater than the peak voltage across the inductor.
b) the peak voltage across the inductor is greater than the peak voltage across the capacitor.
c) the current is in phase with the driving voltage.
d) the peak voltage across the resistor is equal to the peak voltage across the inductor.

Answer: (c)
MC Easy

21.77: The graph, shown, represents resonance if the vertical axis is impedance and the
a) horizontal axis is resistance.
b) straight line is capacitive reactance, the curved line is inductive reactance, and the horizontal axis is frequency.
c) straight line is inductive reactance, the curved line is capacitive reactance, and the horizontal axis is frequency.
d) horizontal axis is inductance.

Answer: (c)
MC   Moderate

21.78: In a given LC resonant circuit,
a) the stored electric field energy is greater than the stored magnetic field energy.
b) the stored electric field energy is less than the stored magnetic field energy.
c) the stored electric field energy is equal to the stored magnetic field energy.
d) all of the above are possible.

Answer: (d)
MC   Moderate

21.79: Consider an RLC series circuit. The impedance of the circuit increases if X_C increases.
a) Always true.
b) True only if X_L is less than or equal to X_C.
c) True only if X_L is greater than or equal to X_C.
d) Never true.

Answer: (b)
MC   Moderate

21.80: Resonance in a series RLC circuit occurs when
a) X_L is greater than X_C.   b) X_C is greater than X_L.
c) $(X_L - X_C)^2$ is equal to R^2.   d) X_C equals X_L

Answer: (d)
MC   Easy

21.81: A resistance of 55 Ω, a capacitor of capacitive reactance 30 Ω, and an inductor of inductive reactance 30 Ω are connected in series to a 110-V, 60-Hz power source. What current flows in this circuit?
a) 2 A   b) Less than 2 A
c) More than 2 A   d) None of the above

Answer: (a)
MC   Moderate

21.82: If the inductance and the capacitance both double in an LRC
series circuit, the resonant frequency of that circuit will
a) decrease to one-half its original value.
b) decrease to one-fourth its original value.
c) decrease to one-eighth its original value.
d) none of the above.

Answer: (a)
MC   Moderate

21.83: A pure capacitor is connected to an AC power supply.   In this
circuit, the current
a) leads the voltage by 90°.
b) lags the voltage by 90°.
c) is in phase with the voltage.
d) None of the above.

Answer: (a)
MC   Easy

21.84: Find the magnetic flux through a level 1 m by 2 m recatangular
table top in a region where the earth's magnetic field is 3.5 x
10^{-5} T and dips downward at a 30° angle below the horizontal.

Answer: 3.5 x 10^{-5} Wb
ES   Easy

21.85: A coil lies flat on a horizontal table top in a region where the
magnetic field points straight down.   The magnetic field
disappears suddenly.   When viewed from above, what is the
direction of the induced current in this coil as the field
disappears?
a) counterclockwise
b) clockwise
c) Clockwise initially, then counterclockwise before stopping.
d) There is no induced current.

Answer: (b)
MC   Moderate

21.86: A coil lies flat on a level table top in a region where the
magnetic field vector points straight up.   The magnetic field
suddenly grows stronger.   When viewed from above, what is the
direction of the induced current in this coil as the field
increases?
a) counterclockwise
b) clockwise
c) Clockwise initially, then counterclockwise before stopping.
d) There is no induced current in this coil.

Answer: (b)
MC   Moderate

21.87: As a coil is removed from a magnetic field an emf is induced in the coil causing a current to flow within the coil. This current interacts with the magnetic field producing a force which
a) acts at right angles to the coil's motion.
b) acts in the direction of the coil's motion.
c) causes the coil to tend to flip over.
d) acts in the direction opposite to the coil's motion.

Answer: (d)
MC   Easy

21.88: A horizontal rod (oriented in the east-west direction) is moved northward at constant velocity through a magnetic field that points straight down. Which statement is true?
a) The west end of the rod is at higher potential than the east end.
b) The east end of the rod is at higher potential than the west end.
c) The top surface of the rod is at higher potential than the bottom surface.
d) The bottom surface of the rod is at higher potential than the top surface.

Answer: (a)
MC   Moderate

21.89: A long straight wire lies on a horizontal table and carries an ever-increasing current northward. Two coils of wire lie flat on the table, one on either side of the wire. When viewed from above, the induced current circles
a) clockwise in both coils.
b) counterclockwise in both coils.
c) clockwise in the east coil and counterclockwise in the west coil.
d) counterclockwise in the east coil and clockwise in the west coil.

Answer: (d)
MC   Moderate

21.90: A circular coil lies flat on a horizontal table. A bar magnet is held above its center with its north pole pointing down, and released. As it approaches the coil, the falling magnet induces (when viewed from above)
a) no current in the coil.
b) a clockwise current in the coil.
c) a counterclockwise current in the coil.
d) a current whose direction cannot be determined from the information provided.

Answer: (c)
MC   Moderate

21.91: A circular coil lies flat on a horizontal table. A bar magnet is held above its center with its north pole pointing down. The stationary magnet induces (when viewed from above)
a) no current in the coil.
b) a clockwise current in the coil.
c) a counterclockwise current in the coil.
d) a current whose direction cannot be determined from the information given

Answer: (a)
MC Moderate

21.92: A circular coil lies flat on a horizontal table. A bar magnet is held above its center with its north pole pointing down, and released. As the magnet falls, the induced current in the coil generates a magnetic field which exerts
a) a constant upward force on the magnet.
b) a constant downward force oon the magnet.
c) an ever-changing upward force on the magnet.
d) an ever-changing downward force on the magnet.

Answer: (c)
MC Moderate

21.93: A 4 mH coil carries a current of 5 A. How much energy is stored in the coil's magnetic field?
a) 0.002 J   b) 0.010 J
c) 0.020 J   d) None of the above.

Answer: (d)
MC Easy

21.94: A series RL circuit with inductance L and resistance R is connected to an emf V. After a period of time, the current reaches a final value of 2 A. A second series circuit is identical except that the inductance is 2L. When it is connected to the same emf V, what will be the final value of the current?
a) 0.5 A   b) 1 A   c) 2 A   d) 4 A

Answer: (c)
MC Moderate

21.95: A coil has a self-inductance of 3 H. What emf is induced in the coil when a 10 A current is reduced to zero in 0.01 s?

Answer: 3000 V
ES Easy

21.96: A resistor and an inductor are connected in series to an ideal battery of constant terminal voltage. At the moment contact is made with the battery, the voltage across the resistor is
   a) greater than the battery's terminal voltage.
   b) equal to the battery's terminal voltage.
   c) less than the battery's terminal voltage, but not zero.
   d) zero.

Answer: (d)
   MC   Moderate

21.97: A resistor and an inductor are connected in series to an ideal battery of constant terminal voltage. At the moment contact is made with the battery, the voltage across the inductor is
   a) greater than the battery's terminal voltage.
   b) equal to the battery's terminal voltage.
   c) less than the battery's terminal voltage, but not zero.
   d) zero.

Answer: (b)
   MC   Moderate

21.98: A simple RL circuit contains a 6 Ω resistor and an 18 H inductor. What is this circuit's time constant?
   a) 108 s   b) 3 s
   c) 0.33 s   d) None of the above.

Answer: (b)
   MC   Easy

21.99: An AC current peaks at 10 A. In passing this current through a resistor, power is dissipated at a particular rate. What DC current through the same resistor would result in the same power dissipation?
   a) 14.1 A   b) 10 A
   c) 7.07 A   d) None of the above.

Answer: (c)
   MC   Easy

21.100: What is the peak voltage in an AC circuit where the rms voltage is 120 V?
   a) 84.8 V   b) 120 V
   c) 170 V   d) None of the above.

Answer: (c)
   MC   Easy

21.101: The potential across an ideal inductor is proportional to
a) the current flowing through it.
b) the resistance of the circuit.
c) the rate of change of current in the coil.
d) None of the above.

Answer: (c)
MC   Easy

21.102: The primary of a transformer has 100 turns and its secondary has
200 turns. If the power input to the primary is 100 W, we can
expect the power output of the secondary to be (neglecting
frictional losses)
a) 50 W   b) 100 W
c) 200 W   d) None of the above.

Answer: (b)
MC   Easy

21.103: The primary of a transformer has 100 turns and its secondary has
200 turns. If the input voltage to the primary is 100 V, we can
expect the output voltage of the secondary to be
a) 50 V   b) 100 V
c) 200 V   d) None of the above.

Answer: (c)
MC   Easy

21.104: The primary of a transformer has 100 turns and its secondary has
200 turns. If the input current at the primary is 100 A, we can
expect the output current at the secondary to be
a) 50 A   b) 100 A
c) 200 A   d) None of the above.

Answer: (a)
MC   Easy

21.105: A transformer has 1000 turns in its primary and 25 turns in its
secondary. The primary is connected to a 120 V, 60 Hz source.
The secondary is connected to a 0.6 Ω resistor. Assuming that
the transformer is 100% efficient, calculate the current that
flows in the primary.

Answer: 0.125 A
ES   Difficult

21.106: A 50 Ω resistor, a 0.25 H coil, and a 1.33×10^{-5} F capacitor are
wired in series to a 120 V, 60 Hz source. What is the voltage
across the resistor?

Answer: 51.5 V
ES   Difficult

21.107: A 50 Ω resistor, a 0.25 H, and a 1.33 x 10⁻⁵ F capacitor are
wired in series to a 120 V, 60 Hz source. What is the voltage
across the capacitor?

Answer: 205 V
ES Difficult

21.108: 100 kW is transmitted down a 6 Ω line at 1000 V. How much less
power would be lost if the power were transmitted at 3000 V
instead of at 1000 V?
a) 1 kW   b) 10 kW   c) 30 kW   d) 53 kW

Answer: (d)
MC Difficult

21.109: In a transformer, how many turns are necessary in a 110 V primary
if the 24 V secondary has 100 turns?
a) 458   b) 240   c) 110   d) 22

Answer: (a)
MC Easy

21.110: 2 A in the 100 turn primary of a transformer causes 14 A to flow
in the secondary. How many turns are in the secondary?
a) 700   b) 114   c) 14   d) 4

Answer: (c)
MC Easy

21.111: 5 A at 110 V flows in the primary of a transformer. Assuming
100% efficiency, how many amps at 24 V can flow in the secondary?
a) 1.1 A   b) 4.6 A   c) 5 A   d) 23 A

Answer: (d)
MC Easy

21.112: What is the inductive reactance of a 2.5 mH coil at 1000 Hz?
a) 2500 Ω   b) 796 Ω   c) 16 Ω   d) 2.5 Ω

Answer: (c)
MC Easy

21.113: What is the current through a 2.5 mH coil due to a 110 V, 60 Hz
source?
a) 0.94 A   b) 2.5 A   c) 104 A   d) 117 A

Answer: (d)
MC Moderate

21.114: At what frequency will a 14 mH coil have 14 Ω of inductive reactance?

    a) 1000 Hz   b) 505 Hz   c) 257 Hz   d) 159 Hz

Answer: (d)
    MC   Easy

21.115: What inductance is necessary for 157 Ω of inductive reactance at 1000 Hz?

    a) 25 mH   b) 157 mH   c) 2.4 H   d) 6.4 H

Answer: (a)
    MC   Easy

21.116: If a 1000 Ω resistor is connected in series with a 20 mH inductor, what is the impedance at 1000 Hz?

    a) 126 Ω   b) 1008 Ω   c) 1126 Ω   d) 126 kΩ

Answer: (b)
    MC   Easy

21.117: If a 1000 Ω resistor is connected in series with a 20 mH inductor, what is the phase angle at 1000 Hz?

    a) 2.9°   b) 7.2°   c) 45°   d) 90°

Answer: (b)
    MC   Moderate

21.118: What inductance is necessary in series with a 500 Ω resistor for a phase angle of 40° at 10 kHz?

    a) 83.9 mH   b) 45.6 mH   c) 20 mH   d) 6.7 mH

Answer: (d)
    MC   Moderate

21.119: What resistance must be put in series with a 450 mH inductor at 5000 Hz for a total impedance of 40000 Ω?

    a) 45400 Ω   b) 40000 Ω   c) 37400 Ω   d) 25800 Ω

Answer: (c)
    MC   Moderate

21.120: What inductance must be put in series with a 100 kΩ resistor at 1 MHz for a total impedance of 150 kΩ?

    a) 17.8 mH   b) 150 mH   c) 167.4 mH   d) 1.5 H

Answer: (a)
    MC   Difficult

21.121: What is the capacitive reactance of a 4.7 μF capacitor at 10 kHz?
a) 0.047 Ω b) 3.4 Ω c) 14.1 Ω d) 47 Ω

Answer: (b)
MC   Easy

21.122: What size capacitor is needed to have 50 Ω of capacitive reactance at 10 kHz?
a) 15.7 μF   b) 5.0 μF   c) 3.2 μF   d) 0.32 μF

Answer: (d)
MC   Easy

21.123: At what frequency will the capacitive reactance of a 0.01 μF capacitor be 100 Ω?
a) 1000 Hz   b) 16 kHz   c) 159 kHz   d) 314 kHz

Answer: (c)
MC   Easy

21.124: What is the current through a 0.001 μF capacitor at 1000 Hz and 5 V?
a) 5.41 μA   b) 31.4 μA   c) 3.14 mA   d) 10.1 mA

Answer: (b)
MC   Moderate

21.125: A 10 Ω resistor is connected in series with a 20 μF capacitor. What is the impedance at 1000 Hz?
a) 8.0 Ω b) 10.0 Ω c) 10.13 Ω d) 12.8 Ω

Answer: (d)
MC   Easy

21.126: A 10 Ω resistor is in series with a 100 μF capacitor at 120 Hz. What is the phase angle?
a) 82.3°   b) 53°   c) 37°   d) 4.7°

Answer: (b)
MC   Moderate

21.127: What resistance is needed in series with a 10 μF capacitor at 1 kHz for a total impedance of 45 Ω?
a) 29.1 Ω b) 42.0 Ω c) 60.9 Ω d) 1772 Ω

Answer: (b)
MC   Moderate

21.128: What resistance is needed in series with a 10 μF capacitor at 1 kHz for a phase angle of 40°?
a) 0.4 Ω　　　b) 2.5 Ω　　　c) 15.7 Ω　　　d) 19 Ω

Answer: (d)
MC　Difficult

21.129: What capacitance is needed in series with a 30 Ω resistor at 1 kHz for a total impedance of 45 Ω?
a) 33.5 μF　　　b) 15 μF　　　c) 4.7 μF　　　d) 0.015 μF

Answer: (c)
MC　Difficult

21.130: What is the total impedance at 1500 Hz if a 100 Ω resistor, 20 mH coil, and 1.0 μF capacitor are connected in series?
a) 189 Ω　　　b) 130 Ω　　　c) 106 Ω　　　d) 82.4 Ω

Answer: (b)
MC　Easy

21.131: What is the phase angle at 1500 Hz if a 100 Ω resistor, 20 mH coil, and 1.0 μF capacitor are connected in series?
a) 0.014°　　　b) 39.5°　　　c) 45°　　　d) 69.5°

Answer: (b)
MC　Easy

21.132: What current flows at 5 V and 1500 Hz if a 100 Ω resistor, 20 mH coil, and 10 μF capacitor are connected in series?
a) 130 mA　　　b) 45 mA　　　c) 39 mA　　　d) 11 mA

Answer: (c)
MC　Moderate

21.133: What resistance is needed in a series circuit with a 20 mH coil and 1.0 μF capacitor for a total impedance of 100 Ω at 1500 Hz?
a) 157 Ω　　　b) 82 Ω　　　c) 57 Ω　　　d) 18 Ω

Answer: (c)
MC　Moderate

21.134: What capacitance is needed in series with a 100 Ω resistor and 15 mH coil to get a total impedance of 110 Ω at 2000 Hz?
a) 143 μF　　　b) 46 μF　　　c) 10 μF　　　d) 0.56 μF

Answer: (d)
MC　Difficult

21.135: What is the resonant frequency of a 1.0 μF capacitor and 15 mH coil in series?
    a) 15 kHz   b) 1300 Hz   c) 770 Hz   d) 67 Hz

Answer: (b)
    MC   Easy

21.136: What inductance is needed in series with a 4.7 μF capacitor for a resonant frequency of 10 kHz?
    a) 21 μH   b) 54 μH   c) 4.7 mH   d) 5.4 mH

Answer: (b)
    MC   Moderate

21.137: What capacitance is needed in series with a 3.7 mH inductor for a resonant frequency of 1000 Hz?
    a) 146 μF   b) 14.6 μF   c) 6.8 μF   d) 0.015 μF

Answer: (c)
    MC   Moderate

21.138: In an AC circuit with a phase angle of 40°, what is the actual power dissipated if the apparent power is 1000 kW?
    a) 1305 kW   b) 1192 kW   c) 839 kW   d) 766 kW

Answer: (d)
    MC   Moderate

21.139: What is the apparent power in an AC circuit if the actual power dissipated is 459 W and the power factor is 0.947?
    a) 435 W   b) 458 W   c) 460 W   d) 485 W

Answer: (d)
    MC   Easy

21.140: The phase angle of an AC circuit is 63°. What is the power factor?
    a) 0.891   b) 0.546   c) 0.454   d) 0.109

Answer: (c)
    MC   Easy

22. Electromagnetic Waves

22.1: **MATCH THE RADIATION TO ITS TYPICAL WAVELENGTH.**

22.1: visible light

Answer: 10^{-7} m
MA   Easy   Table: 1

22.2: microwaves

Answer: centimeters
MA   Easy   Table: 1

22.3: x-rays

Answer: 10^{-11} m
MA   Easy   Table: 1

22.4: TV waves

Answer: meters
MA   Easy   Table: 1

22.5: infrared

Answer: 10^{-5} m
MA   Easy   Table: 1

22.6: All electromagnetic waves travel through a vacuum at
   a) the same speed.
   b) speeds that are proportional to their frequency.
   c) speeds that are inversely proportional to their frequency.
   d) None of the above.

Answer: (a)
   MC   Easy

22.7: Electromagnetic waves are
   a) longitudinal.
   b) transverse.
   c) both longitudinal and transverse.
   d) None of the above.

Answer: (b)
   MC   Easy

22.8: The **E** and **B** fields in electromagnetic waves are oriented
a) parallel to the wave's direction of travel, as well as to each other.
b) parallel to the waves direction of travel, and perpendicular to each other.
c) perpendicular to the wave's direction of travel, and parallel to each other.
d) perpendicular to the wave's direction of travel, and also to each other.

Answer: (d)
MC Moderate

22.9: An electromagnetic wave is radiated by a straight wire antenna that is oriented vertically. What should be the orientation of a straight wire receiving antenna? It should be placed
a) vertically.
b) horizontally and in a direction parallel to the wave's direction of motion.
c) horizontally and in a direction perpendicular to the wave's direction of motion.
d) None of the above.

Answer: (a)
MC Moderate

22.10: Electromagnetic waves can travel through
a) glass.   b) iron.
c) water.   d) All of the above.

Answer: (d)
MC Easy

22.11: In a vacuum, the velocity of all electromagnetic waves
a) is zero.
b) is nearly 3×10^8 m/s.
c) depends on the frequency.
d) depends on their amplitude.

Answer: (b)
MC Easy

22.12: An electromagnetic wave is traveling to the east. At one instant at a given point its **E** vector points straight up. What is the direction of its **B** vector?
a) north   b) down   c) east   d) south

Answer: (d)
MC Moderate

22.13: Of the following, which are not electromagnetic in nature?
      a) microwaves   b) gamma rays
      c) sound waves   d) radio waves

Answer: (c)
     MC   Easy

22.14: Which of the following correctly lists electromagnetic waves in order from longest to shortest wavelength?
      a) gamma rays, ultraviolet, infrared, microwaves
      b) microwaves, ultraviolet, visible light, gamma rays
      c) radio waves, infrared, gamma rays, ultraviolet
      d) television, infrared, visible light, X-rays

Answer: (d)
     MC Moderate

22.15: What is the wavelength of light waves if their frequency is 5 x 10^{14} Hz?
      a) 0.6 m   b) 6 mm   c) 0.06 mm   d) 0.6 μm

Answer: (d)
     MC Easy

22.16: How far is a light year (the distance light travels in a year)?
      a) 186,000 m   b) 3 x 10^8 m
      c) 8.7 x 10^{13} m   d) 9.5 x 10^{15} m

Answer: (d)
     MC Moderate

22.17: How far does light travel in 1 μs?
      a) 3 x 10^{14} m   b) 300 m   c) 3 m   d) 30 cm

Answer: (b)
     MC Easy

22.18: How long does it take light to travel 1 m?
      a) 3.3 ns   b) 3.3 μs   c) 3.3 ms   d) 3.3 s

Answer: (a)
     MC Easy

22.19: How long does it take the signal from a local radio station to travel 10 miles?
      a) 1.4 s   b) 53.8 ms   c) 18.6 ms   d) 53.8 μs

Answer: (d)
     MC Easy

22.20: What is the wavelength of a 92.9 MHz radio wave?
   a) 32 mm   b) 32 cm   c) 3.2 m   d) 32 m

Answer: (c)
   MC   Easy

22.21: What frequency are 20 mm microwaves?
   a) 100 MHz   b) 400 MHz   c) 15 GHz   d) 73 GHz

Answer: (c)
   MC   Easy

22.22: What is the frequency of 10 m radio waves?
   a) 500 kHz   b) 1.7 MHz   c) 30 MHz   d) 30 GHz

Answer: (c)
   MC   Easy

23. Light: Geometric Optics

23.1: An object is placed 15 cm from a concave mirror of focal length 20 cm. The object is 4 cm tall. How tall is the image, and where is it located?

Answer: 16 cm tall, 60 cm behind the mirror.
ES   Moderate

23.2: An object is located 2 m in front of a plane mirror. The image formed by the mirror appears to be
   a) 1 m in front of the mirror.
   b) on the mirror's surface.
   c) 1 m behind the mirror's surface.
   d) 2 m behind the mirror's surface.

Answer: (d)
MC   Easy

23.3: A spherical concave mirror has a radius of curvature of 50 cm. How far from the mirror is the focal point located?
   a) 25 cm   b) 50 cm   c) 75 cm   d) 100 cm

Answer: (a)
MC   Easy

23.4: A light ray, traveling parallel to a concave mirror's axis, strikes the mirror's surface near its midpoint. After reflection, this ray
   a) again travels parallel to the mirror's axis.
   b) travels at right angles to the mirror's axis.
   c) passes through the mirror's center of curvature.
   d) passes through the mirror's focal point.

Answer: (d)
MC   Moderate

23.5: A light ray, traveling obliquely to a concave mirror's axis, crosses the axis at the mirror's center of curvature before striking the mirror's surface. After reflection, this ray
   a) travels parallel to the mirror's axis.
   b) travels at right angles to the mirror's axis.
   c) passes through the mirror's center of curvature.
   d) passes through the mirror's focal point.

Answer: (c)
MC   Moderate

23.6: A light ray, traveling obliquely to a concave mirror's surface, crosses the axis at the mirror's focal point before striking the mirror's surface. After reflection, this ray
   a) travels parallel to the mirror's axis.
   b) travels at right angles to the mirror's axis.
   c) passes through the mirror's center of curvature.
   d) passes through the mirror's focal point.

Answer: (a)
   MC   Moderate

23.7: An object is placed at a concave mirror's center of curvature. The image produced by the mirror is located
   a) out beyond the center of curvature.
   b) at the center of curvature.
   c) between the center of curvature and the focal point.
   d) at the focal point.

Answer: (b)
   MC   Moderate

23.8: An object is positioned between a concave mirror's center of curvature and its focal point. The image produced by the mirror is located
   a) out past the center of curvature.
   b) at the center of curvature.
   c) between the center of curvature and the focal point.
   d) at the focal point.

Answer: (a)
   MC   Moderate

23.9: An object is situated between a concave mirror's surface and its focal point. The image formed in this case is
   a) real and inverted.   b) real and erect.
   c) virtual and erect.   d) virtual and inverted.

Answer: (c)
   MC   Moderate

23.10: A single concave spherical mirror produces an image which is
   a) always virtual.
   b) always real.
   c) real only if the object distance is less than f.
   d) real only if the object distance is greater than f.

Answer: (d)
   MC   Moderate

23.11: A single convex spherical mirror produces an image which is
a) always virtual.
b) always real.
c) real only if the object distance is less than f.
d) real only if the object distance is greater than f.

Answer: (a)
MC   Moderate

23.12: Concave spherical mirrors produce images which
a) are always smaller than the actual object.
b) are always larger than the actual object.
c) are always the same size as the actual object.
d) could be smaller than, larger than, or the same size as the actual object, depending on the placement of the object.

Answer: (d)
MC   Moderate

23.13: Convex spherical mirrors produce images which
a) are always smaller than the actual object.
b) are always larger than the actual object.
c) are always the same size as the actual object.
d) could be larger than, smaller than, or the same size as the actual object, depending on the placement of the object.

Answer: (a)
MC   Moderate

23.14: Plane mirrors produce images which
a) are always smaller than the actual object.
b) are always larger than the actual object.
c) are always the same size as the actual object.
d) could be smaller, larger, or the same size as the actual object, depending on the placement of the object.

Answer: (c)
MC   Easy

23.15: A light ray, traveling parallel to the axis of a convex thin lens, strikes the lens near its midpoint. After traveling through the lens, this ray emerges traveling obliquely to the axis of the lens
a) such that it never crosses the axis again
b) crossing the axis at a point equal to twice the focal length.
c) passing between the lens and its focal point.
d) passing through its focal point.

Answer: (d)
MC   Moderate

23.16: A light ray, traveling parallel to the axis of a thin concave lens, strikes the lens near its midpoint. After traveling though the lens, this ray emerges traveling obliquely to the axis of the lens
a) such that it never crosses the axis.
b) crossing the axis at a point equal to twice the focal length.
c) passing between the lens and its focal point.
d) passing through the focal point.

Answer: (a)
MC   Moderate

23.17: A convex lens has focal length f. An object is placed at 2f on the axis. The image formed is located
a) at 2f.   b) between f and 2f.
c) at f.   d) between the lens and f.

Answer: (a)
MC   Easy

23.18: A convex lens has a focal length f. An object is placed between f and 2f on the axis. The image formed is located
a) at 2f
b) between f and 2f.
c) at f.
d) at a distance greater than 2f from the lens.

Answer: (d)
MC   Moderate

23.19: A convex lens has a focal length f. An object is placed between infinity and 2f from the lens on its axis. The image formed is located
a) at 2f.   b) between f and 2f.
c) at f.   d) between the lens and f.

Answer: (b)
MC   Moderate

23.20: A convex lens has focal length f. An object is located at infinity. The image formed is located
a) at 2f.   b) between f and 2f.
c) at f.   d) between the lens and f.

Answer: (c)
MC   Easy

23.21: A convex lens has a focal length f. An object is placed at f on the axis. The image formed is located
a) at infinity.   b) between 2f and infinity.
c) at 2f.   d) between f and 2f.

Answer: (a)
MC   Easy

23.22: A object is placed between a convex lens and its focal point. The image formed is
a) virtual and erect.   b) virtual and inverted.
c) real and erect.   d) real and inverted.

Answer: (a)
MC   Moderate

23.23: A concave spherical mirror has a focal length of 20 cm. An object is placed 30 cm in front of the mirror on the mirror's axis. Where is the image located?
a) 12 cm in front of the mirror.
b) 60 cm behind the mirror.
c) 60 cm in front of the mirror.
d) None of the above.

Answer: (c)
MC   Easy

23.24: A convex spherical mirror has a focal length of -20 cm. An object is placed 30 cm in front of the mirror on the mirror's axis. Where is the image located?
a) 12 cm in front of the mirror.
b) 60 cm behind the mirror.
c) 60 cm in front of the mirror.
d) None of the above.

Answer: (d)
MC   Easy

23.25: A concave spherical mirror has a focal length of 20 cm. An object is placed 10 cm in front of the mirror on the mirror's axis. Where is the image located?
a) 20 cm behind the mirror.
b) 20 cm in front of the mirror.
c) 6.67 cm behind the mirror.
d) 6.67 cm in front of the mirror.

Answer: (a)
MC   Easy

23.26: A convex spherical mirror has a focal length of -20 cm. An object is placed 10 cm in front of the mirror on the mirror's axis. Where is the image located?
a) 20 cm behind the mirror.
b) 20 cm in front of the mirror.
c) 6.67 cm behind the mirror.
d) 6.67 cm in front of the mirror.

Answer: (c)
MC   Easy

23.27: How far are you from your image when you stand 0.75 m in front of
a vertical plane mirror?
a) 0.75 m
b) 1.5 m
c) 3.0 m
d) None of the above.

Answer: (b)
MC   Easy

23.28: How fast do you approach your image when you approach a vertical
plane mirror at a speed of 2 m/s?
a) 1 m/s
b) 2 m/s
c) 4 m/s
d) None of the above.

Answer: (c)
MC   Easy

23.29: An object is placed 40 cm in front of a 20 cm focal length
converging lens. How far is the image of this object from the
lens?
a) 40 cm
b) 20 cm
c) 13.3 cm
d) None of the above.

Answer: (a)
MC   Easy

23.30: You hold a square mirror (15 cm on a side) 40 cm from your eyes
and see a building directly behind you. If the building appears
to fill the vertical height of the mirror, and the building is
located 100 m behind you, how tall is the building?

Answer: 37.6 m
ES   Moderate

23.31: A 5 cm tall object is placed 60 cm in front of a concave
spherical mirror of focal length 20 cm. What is the height of
the image?

Answer: 2.5 cm
ES   Moderate

23.32: A 5 cm tall object is placed 60 cm in front of a convex spherical
mirror of focal length 20 cm. What is the height of the image?

Answer: 1.25 cm
ES   Moderate

23.33: A medium that almost all light passes through is called
a) transparent.
b) translucent.
c) opaque.
d) diffuse.

Answer: (a)
MC   Easy

23.34: The angle of incidence of a light ray is equal to
a) 45°   b) the angle of the image.
c) the angle of reflection.   d) the normal.

Answer: (c)
MC   Easy

23.35: The laws of reflection apply only to plane mirrors.

Answer: False
TF   Easy

23.36: A plane mirror will form what kind of image(s)?
a) multiple   b) real   c) inverted   d) virtual

Answer: (d)
MC   Easy

23.37: If the radius of curvature of the concave mirror is r, the focal length is
a) 2r
b) r
c) r/2
d) Cannot be determined from the information given.

Answer: (c)
MC   Easy

23.38: Light arriving at a concave mirror on a path parallel to the axis is reflected
a) back parallel to the axis.
b) back on itself.
c) through the focal point.
d) through the center of curvature.

Answer: (c)
MC   Moderate

23.39: Light arriving at a concave mirror on a path through the focal point is reflected
a) back parallel to the axis.
b) back on itself.
c) through the focal point.
d) through the center of curvature.

Answer: (a)
MC   Moderate

23.40: An object is 10 cm in front of a concave mirror with focal length 3 cm. Where is the image?
a) 13 cm from the mirror
b) 7 cm from the mirror
c) 4.3 cm from the mirror
d) 3.3 cm from the mirror

Answer: (c)
MC   Easy

23.41: A object is 12 cm in front of a concave mirror, and the image is 3 cm in front of the mirror. What is the focal length of the mirror?
a) 15 cm
b) 7.9 cm
c) 2.4 cm
d) 1.3 cm

Answer: (c)
MC   Easy

23.42: An image is 4 cm behind a concave mirror with focal length 5 cm. Whare is the object?
a) 2.2 cm in front of the mirror.
b) 2.2 cm behind the mirror.
c) 9.0 cm in front of the mirror.
d) 1.0 cm behind the mirror.

Answer: (a)
MC   Moderate

23.43: A 1.4 cm tall object is 4.0 cm from a concave mirror. If the image is 4.0 cm tall, how far is it from the mirror?
a) 11.4 cm
b) 9.4 cm
c) 1.4 cm
d) 0.09 cm

Answer: (a)
MC   Easy

23.44: An object is 14 cm in front of a convex mirror. The image is 5.8 cm behind the mirror. What is the focal length of the mirror?
a) -4.1 cm
b) -8.2 cm
c) -9.9 cm
d) -19.8 cm

Answer: (c)
MC   Moderate

23.45: An object is 5.7 cm from a concave mirror. The image is 4.7 cm tall, and 10 cm from the mirror. How tall is the object?
a) 12.1 cm
b) 11.1 cm
c) 8.2 cm
d) 2.7 cm

Answer: (d)
MC   Moderate

23.46: An object is 47.5 cm tall. The image is 38.6 cm tall, and 14.8 cm from the mirror. How far is the object from the mirror?
a) 123.9 cm
b) 47.6 cm
c) 18.2 cm
d) 12.0 cm

Answer: (c)
MC   Moderate

23.47: An object is 8.9 cm tall. The image is 7.8 cm tall, and 14.8 cm from a convex mirror. What is the mirror's focal length?
a) -120 cm   b) -105 cm   c) -16.9 cm   d) -13.0 cm

Answer: (a)
MC Difficult

23.48: Light traveling at an angle into a denser medium is refracted
a) toward the normal.   b) away from the normal.
c) parallel to the normal.   d) equally.

Answer: (a)
MC Easy

23.49: Light passes from air to water. The incoming ray is at an angle of 17° to the normal. The index of refraction is 1.33. What is the angle in the water?
a) 22.9°   b) 22.6°   c) 18.3°   d) 12.7°

Answer: (d)
MC Easy

23.50: Light enters a substance from air at an angle of 32°, and continues at an angle of 23°. What is the index of refraction of the substance?
a) 0.74   b) 1.11   c) 1.28   d) 1.36

Answer: (d)
MC Easy

23.51: Lucite has an index of refraction of 1.50. How fast will light travel through it?
a) 2.0×10^8 m/s   b) 3.0×10^8 m/s
c) 4.5×10^8 m/s   d) 6.0×10^8 m/s

Answer: (a)
MC Easy

23.52: Lucite has an index of refraction of 1.50. What is its critical angle of incidence?
a) 1.16°   b) 15°   c) 41.8°   d) 87.4°

Answer: (c)
MC Easy

23.53: Light enters a substance from air at 30° to the normal. It continues through the substance at 23° to the normal. What would be the critical angle for this substance?
a) 53°   b) 51.4°   c) 36.7°   d) 12.6°

Answer: (b)
MC Moderate

23.54: The critical angle for a substance is measured at 53.7°. Light enters from air at 45°. At what angle will it continue?
   a) 34.7°
   b) 45°
   c) 53.7°
   d) It will not continue, but be totally reflected.

Answer: (a)
MC   Moderate

23.55: A substance has an index of refraction of 1.46. Light is passing through it at 53°. At what angle will it leave into the air?
   a) It will not leave.   b) 59.1°
   c) 43.2°   d) 33.2°

Answer: (a)
MC   Easy

23.56: Lenses that are thicker at the center
   a) spread out light rays.
   b) bend light rays to a point beyond the lens.
   c) have no effect on light rays.
   d) refelect light rays back.

Answer: (b)
MC   Easy

23.57: An object is 12 cm in front of a converging lens with focal length 4 cm. Where is the image?
   a) 8 cm behind the lens   b) 6 cm in front of the lens
   c) 6 cm behind the lens   d) 4 cm in front of the lens

Answer: (c)
MC   Easy

23.58: An object is 15 mm in front of a converging lens, and the image is 4 mm behind the lens. What is the focal length of the lens?
   a) 11 mm   b) 5.5 mm   c) 3.8 mm   d) 3.2 mm

Answer: (d)
MC   Easy

23.59: An image is 4 mm in front of a converging lens with focal length 5 mm. Where is the object?
   a) 2.2 mm in front of the lens
   b) 2.2 mm behind the lens
   c) 9 mm behind the lens
   d) 20 mm in front of the lens

Answer: (a)
MC   Moderate

23.60: A 14 mm tall object is 4 mm from a converging lens. If the image is 4 mm tall, how far is it from the lens?
    a) 14 mm   b) 8.7 mm   c) 1.4 mm   d) 1.1 mm

Answer: (d)
    MC   Easy

23.61: An object is 1.4 mm in front of a converging lens. The image is 5.8 mm behind the lens. What is the focal length of the lens?
    a) 4.4 mm   b) 4.1 mm   c) 1.4 mm   d) 1.1 mm

Answer: (d)
    MC   Easy

23.62: An object is 15.2 mm from a converging lens. The image is 4 mm tall, and 9 cm from the lens. How tall is the object?
    a) 6.8 mm   b) 5.4 mm   c) 1.7 mm   d) 0.68 mm

Answer: (d)
    MC   Moderate

23.63: An object is 4.1 cm tall, and 10.3 cm from a converging lens. The image is virtual and 6.2 cm tall. What is the focal length of the lens?
    a) 6.8 cm   b) 10.3 cm   c) 15.6 cm   d) 30.4 cm

Answer: (d)
    MC   Difficult

23.64: An object is 6 cm tall, and is in front of a diverging lens. The image is 2.5 cm tall, and 7.5 cm from the lens. What is the focal length of the lens?
    a) -6 cm   b) -7.5 cm   c) -12.9 cm   d) -18 cm

Answer: (c)
    MC   Difficult

23.65: An object is 10.4 cm tall, and 4.8 cm in front of a diverging lens. The image is 4 cm from the lens. How tall is the image?
    a) 12.5 cm   b) 8.7 cm   c) 5.4 cm   d) 1.8 cm

Answer: (b)
    MC   Moderate

23.1: **MATCH THE RELATIONSHIP TO THE PHYSICAL PHENOMENON.**

23.66: Snell's law

Answer: $(n_1 \sin\theta_1 / n_2 \sin\theta_2) = 1$
    MA   Easy

23.67: critical angle

Answer: arc sin(n_2/n_1)
         MA   Easy

23.68: index of refraction

Answer: c/v
         MA   Easy

23.69: Law of reflection

Answer: $\theta_i = \theta_r$
         MA   Easy

23.70: A laser beam strikes a plane's reflecting surface with an angle
        of incidence of 37°. What is the angle between the incident ray
        and the reflected ray?

Answer: 74°
         ES   Easy

23.71: An index of refraction less than one for a medium would imply
        a) that the speed of light in the medium is the same as the
        speed of light in air.
        b) that the speed of light in the medium is greater than the
        speed of light in air.
        c) refraction is not possible.
        d) reflection is not possible.

Answer: (b)
         MC   Easy

23.72: A light beam enters water at an angle of incidence of 37°.
        Determine the angle of refraction.

Answer: 26.9°
         ES   Easy

23.73: A buoy, used to mark a harbor channel, consists of a weighted rod
        2 m in length. 1.5 m of the rod is immersed in water and 0.5 m
        extends above the surface. If sunlight is incident on the water
        with an angle of incidence of 40°, what is the length of the
        shadow of the buoy on the level bottom of the harbor?

Answer: 1.25 m
         ES   Difficult

23.74: A light ray strikes a glass plate of thickness 0.8 cm at an angle of incidence of 60°. The index of refraction of the glass is 1.55. By how much is the beam displaced from its original line of travel after it has passed through the glass?

Answer: 0.42 cm
ES Difficult

23.75: Light travels fastest
a) in a vacuum. b) through water.
c) through glass. d) through diamond.

Answer: (a)
MC Easy

23.76: A ray of light, which is traveling in air, is incident on a glass plate at a 45° angle. The angle of refraction in the glass
a) is less than 45°.
b) is greater than 45°.
c) is equal to 45°.
d) could be any of the above; it all depends on the index of refraction of glass.

Answer: (a)
MC Easy

23.77:

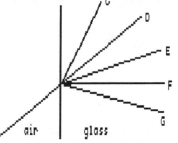

Shown here are some possible paths that a light ray can follow in going either from glass to air or from air into glass. In the drawing you are not told whether the light is going from left to right or from right to left. Which path did the light follow?
a) One cannot answer this question without knowing in which direction the light is going.
b) More than one of these paths is possible.
c) Path C
d) Path D
e) Path E
f) Path F
g) Path G

Answer: (e)
MC Easy

324

23.78: A ray of light, which is traveling in a vacuum, is incident on a glass plate (n = ng). For increasing angles of incidence, the angle of refraction
a) increases, approaching the limiting value of 90°.
b) increases, approaching the limiting value of $\sin^{-1}(1/n_g)$ degrees.
c) increases, approaching the limiting value of $\sin^{-1}(n_g)$ degrees.
d) decreases, approaching the limiting value of zero degrees.

Answer: (b)
MC   Moderate

23.79: For all transparent material substances, the index of refraction
a) is less than 1.
b) is greater than 1.
c) is equal to 1.
d) could be any of the above; it all depends on optical density.

Answer: (b)
MC   Easy

23.80: The angle of incidence
a) must equal the angle of refraction.
b) is always less than the angle of refraction.
c) is always greater than the angle of refraction.
d) may be greater than, less than, or equal to the angle of refraction.

Answer: (d)
MC   Easy

23.81: The angle of incidence
a) must equal the angle of reflection.
b) is always less than the angle of reflection.
c) is always greater than the angle of reflection.
d) may be greater than, less than, or equal to the angle of reflection.

Answer: (a)
MC   Easy

23.82: An oil layer that is 5 cm thick is spread smoothly and evenly over the surface of water on a windless day. What is the angle of refraction in the water for a ray of light that has an angle of incidence of 45° as it enters the oil from the air above? (The index of refraction for oil is 1.15, and for water it is 1.33.)
a) 27.2°   b) 32.1°   c) 35.5°   d) 38.6°

Answer: (b)
MC   Moderate

23.83: A beam of light, traveling in air, strikes a plate of transparent material at an angle of incidence of 56°. It is observed that the reflected and refracted beams form an angle of 90°. What is the index of refraction of this material?
   a) n = 1.40   b) n = 1.43   c) n = 1.44   d) n = 1.48

Answer: (d)
      MC   Easy

23.84: A point source of light is positioned 20 m below the surface of a lake. What is the diameter of the largest circle on the surface of the water through which light can emerge?

Answer: 45.6 m
      ES   Moderate

23.85: The end of a cylindrical plastic rod is polished and cut perpendicular to the axis of the cylinder. Determine the minimum index of refraction so that a light ray entering the end of the rod will always be totally internally reflected within the rod, i.e., it will never escape the rod until it comes to the other end.

Answer: 1.414
      ES   Difficult

23.86: The critical angle for a beam of light passing from water into air is 48.8°. This means that all light rays with an angle of incidence greater than this angle will be
   a) absorbed.
   b) totally reflected.
   c) partially reflected and partially transmitted.
   d) totally transmitted.

Answer: (b)
      MC   Moderate

23.87: What is the critical angle for light traveling from crown glass (n = 1.52) into water (n = 1.33)?
   a) 42°   b) 48°   c) 57°   d) 61°

Answer: (d)
      MC   Easy

23.88: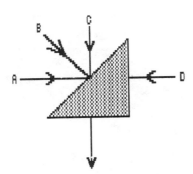

A light ray emerges from the bottom of a 45° glass prism into air, as shown in the sketch. From which direction was the ray incident on the prism?

a) A b) B c) C d) D

Answer: (d)
MC Moderate

23.89: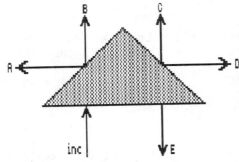

A ray of light (inc) enters a 45° - 90° glass prism from air (bottom left), as shown. In what direction does the light re-emerge back into the air?

a) A b) B c) C d) D e) E

Answer: (e)
MC Moderate

23.90: An optical fiber is 1 meter long and has a diameter of 20 μm. Its ends are perpendicular to its axis. Its index of refraction is 1.30. What is the maximum number of reflections a light ray entering one end will make before it emerges from the other end?

a) 28,500 b) 25,600 c) 24,220 d) 18,500

Answer: (d)
MC Difficult

23.91: Endoscopes are based on the physical phenomena called
a) absorption. b) dispersion.
c) total internal reflection. d) diffusion.

Answer: (c)
MC Moderate

23.92: Magnification

Answer: h_i/h_o
MA Easy Table: 1

23.93: Mirror equation

Answer: $1/f = 1/d_o + 1/d_i$
MA   Easy   Table: 1

23.94: Lens maker's equation

Answer: $1/f = (n-1) (1/R_1 - 1/R_2)$
MA   Easy   Table: 1

23.95: A plane mirror forms an image that is
   a) real and upright.   b) virtual and upright.
   c) real and upside down.   d) virtual and upside down.

Answer: (b)
   MC   Easy

23.96: If you place a plane mirror vertically into water, and a dolphin
   moves toward it with a speed of 4 m/s, she will see her image
   move toward her at 4 m/s.

Answer: False
   TF   Easy

23.97: You may have seen ambulances on the street with the letters of
   the word AMBULANCE written on the front of them, in such a way as
   to appear correctly when viewed in your car's rear-view mirror.
   How do the letters appear when you look directly at the
   ambulance?
   a)
      AMBULANCE   b)   ECNALUBMA

   c)   ECNALUBMA   d)   AMBULANCE

   e)   AMBULANCE

Answer: (e)
   MC   Easy

23.98: The simplest type of camera is a pin-hole camera; you might have
   even made one. It consists of a box with a single pinhole in it,
   and a piece of film on the inner side opposite the hole. Suppose
   you photograph your friend with it. Your friend's image on the
   film will appear
   a) upside-down because light travels in a straight line.
   b) right-side up because light travels in a straight line.
   c) upside-down because light refracts at the pinhole.
   d) right-side up because light refracts at the pinhole.

Answer: (a)
   MC   Moderate

23.99: A man 190 cm tall stands 2.5 m from a plane wall mirror. One meter in front of him sits his dog, 30 cm tall. What minimum mirror size is needed for the man to see all of his dog's reflection?

Answer: 18.8 cm tall
ES Moderate

23.100: A friend who is 180 cm tall stands 2 m from a plane wall mirror. You stand slightly behind him, 3 m from the wall mirror. How tall must the mirror be if you are to see all of your friend in the mirror?

Answer: 1.08 m
ES Moderate

23.101: 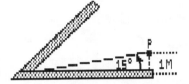 At an amusement park you stand between two plane mirrors, which intersect at an angle of 60°. You are positioned at point P in the drawing here. You are 1 m from the nearest mirror, and 15° above it. What is the distance from you to the nearest image and to the second nearest image?

Answer: 2 m; 5.5 m
ES Difficult

23.102: Two plane mirrors intersect at an angle of 120°. Draw ray diagrams to determine the locations of all the images formed by an object placed at a point P between the mirrors, closer to one mirror than the other.

Answer: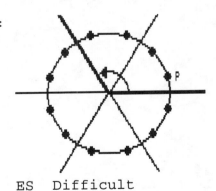

ES Difficult

23.103: I recently had to replace the side view mirror on my truck. How tall should the mirror have been if I am to see all of a car 2 m high when it is following 20 m back (measured from the mirror). The mirror is 0.5 m from my eyes.
  a) 4.9 cm   b) 5.2 cm   c) 5.8 cm   d) 7.0 cm

Answer: (a)
MC Moderate

23.104: An object is placed 6 cm from a concave mirror of radius 4 cm. Graphically determine (a) the image position. (b) if the image is larger, smaller, or the same size as the object. (c) if the image is real or virtual.

Answer: (a) (shown)
       (b) smaller
       (c) real

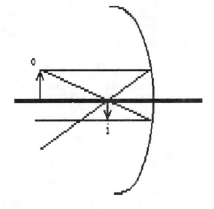

ES   Moderate

23.105:

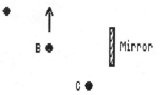

Suppose a lighted candle is placed a short distance from a plane mirror, as shown here. Where will the image of the flame be located?
   a) At A   b) At B
   c) At C   d) At M (at the mirror)

Answer: (c)
MC   Easy

23.106:

At which, if any, of the points indicated here should you place your eye if you wish to see an image of the arrow in the mirror?
   a) A   b) B
   c) C   d) D
   e) None of the above

Answer: (d)
MC   Easy

23.107: An object is placed a distance d in front of a plane mirror. The size of the image will be
   a) half as big as the size of the object.
   b) dependent on the distance d.
   c) dependent on where you are positioned when you look at the image.
   d) twice the size of the object.
   e) the same size as the object, independent of the distance d or the position of the observer.

Answer: (e)
   MC   Easy

23.108: The rear-view mirrors on the passenger side of many new cars have a warning written on them: "OBJECTS IN MIRROR ARE CLOSER THAN THEY APPEAR." This implies that the mirror must be concave.

Answer: True
   TF   Moderate

23.109: If you stand in front of a concave mirror, exactly at its focal point,
   a) you won't see your image because there is none.
   b) you won't see your image because it's focused at a different distance.
   c) you will see your image, and you will appear smaller.
   d) you will see your image and you will appear larger.
   e) you will see your image at your same height.

Answer: (a)
   MC   Moderate

23.110: If you stand in front of a concave mirror, exactly at its center of curvature,
   a) you won't see your image because there is none.
   b) you won't see your image because it's focused at a different distance.
   c) you will see your image and you will appear smaller.
   d) you will see your image and you will appear larger.
   e) you will see your image at your same height.

Answer: (e)
   MC   Moderate

23.111: If you stand in front of a convex mirror, at the same distance from it as its focal length,
   a) you won't see your image because there is none.
   b) you won't see your image because it's focused at a different distance.
   c) you will see your image and you will appear smaller.
   d) you will see your image and you will appear larger.
   e) you will see your image at your same height.

Answer: (a)
   MC   Moderate

23.112: If you stand in front of a convex mirror, at the same distance from it as its radius of curvature,
   a) you won't see your image because there is none.
   b) you won't see your image because it's focused at a different distance.
   c) you will see your image and you will appear smaller.
   d) you will see your image and you will appear larger.
   e) you will see your image at your same height.

Answer: (c)
   MC   Moderate

23.113: An object 2 cm tall is placed 24 cm in front of a convex mirror whose focal length is 30 cm. (a) Where is the image formed?   (b) How tall is it?

Answer: (a) 13.2 cm behind the mirror   (b) 1.1 cm tall
   ES   Moderate

23.114: An object is placed 15 cm from a concave mirror of focal length 20 cm. If the object is 4 cm tall, (a) where is the image formed?   (b) how tall is it?

Answer: (a) 60 cm behind the mirror   (b) 16 cm tall
   ES   Moderate

23.115: In using ray tracing to graphically locate the image of an object that is placed in front of a mirror, describe three simple rays that you could draw, that pass by the head of the object.

Answer: Ray 1: the radial ray; i.e., the ray drawn toward the center of curvature, which is then reflected back along the same path.
   Ray 2: the parallel ray; i.e., the incoming ray parallel to the main axis, which is then reflected along a path that extends to the focal point.
   Ray 3: the focal ray; i.e., the ray drawn toward the focal point, which is then reflected back parallel to the main axis.
   ES   Moderate

23.116: An object is placed 3 cm from a convex mirror of radius 4 cm.
(a) Graphically determine the size and position of the image.
(b) Is the image real or virtual?

Answer: (a) (see sketch)
(b) virtual

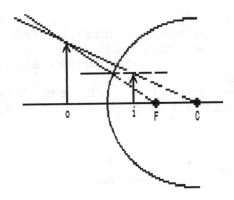

ES   Moderate

23.117: Sometimes when you look into a curved mirror you see a magnified image (a great big you!) and sometimes you see a diminished image (a little you). If you look at the bottom (convex) side of a shiny spoon, what will you see?
a) You won't see an image of yourself because no image will be formed.
b) You will see a little you, upside down.
c) You will see a little you, right side up.
d) You will see a little you, but whether you are right side up or upside down depends on how near you are to the spoon.
e) You will either see a little you or a great big you, depending on how near you are to the spoon.

Answer: (c)
MC   Moderate

23.118: Which of the following is an accurate statement?
a) A mirror always forms a real image.
b) A mirror always forms a virtual image.
c) A mirror always forms an image larger than the object.
d) A mirror always forms an image smaller than the object.
e) None of the above is true.

Answer: (e)
MC   Moderate

23.119: A negative magnification for a mirror means
a) the image is inverted, and the mirror is concave.
b) the image is inverted, and the mirror is convex.
c) the image is inverted, and the mirror may be concave or convex.
d) the image is upright, and the mirror is convex.
e) the image is upright, and the mirror may be concave or convex.

Answer: (c)
MC   Moderate

23.120: In using ray tracing to graphically locate the image of an object that is placed in front of a lens, describe three simple rays that you could draw that pass by the head of the object.

Answer: Ray 1: the vertex ray; i.e., the ray passing through the center of the lens, which emerges undeflected.
Ray 2: the parallel ray; i.e., the incoming ray parallel to the main axis, which is then refracted along a path that extends to the focal point.
Ray 3: the focal ray; i.e., the ray drawn toward the focal point, which is then refracted parallel to the main axis.
ES Moderate

23.121: An object is placed 9.5 cm from a lens of focal length 24 cm. (a) Where is the image formed? (b) What is the magnification?

Answer: (a) 15.7 cm from the lens on the same side as the object. (b) m = 1.7
ES Moderate

23.122: The image of the rare stamp you see through a magnifying glass is
a) always the same orientation as the stamp.
b) always upside-down compared to the stamp.
c) either the same orientation or upside-down, depending on how close the stamp is to the glass.
d) either the same orientation or upside-down, depending on the thickness of the glass used.

Answer: (c)
MC Moderate

23.123: Where must an object be placed with respect to a converging lens of focal length 30 cm if the image is to be virtual, and three times as large as the object?

Answer: 20 cm from the lens
ES Difficult

23.124: A camera with a telephoto lens of focal length 125 mm is used to take a photograph of a plant 1.8 m tall. The plant is 5.0 m from the lens. (a) What must be the distance between the lens and the camera film if the image is sharply focused? (b) How tall is the image?

Answer: (a) 128 mm from lens to film   (b) 46 mm tall
ES Moderate

23.125: A diverging lens is thicker at its center than it is at its edges.

Answer: False
TF Easy

23.126: The images formed by concave lenses
  a) are always real.
  b) are always virtual.
  c) could be real or virtual; it depends on whether the object
  distance is smaller or greater than the focal length.
  d) could be real or virtual, but always real when the object is
  placed at the focal point.

Answer: (b)
  MC Easy

23.127: A slide projector has a lens of focal length 15 cm. An image 1 m
  x 1 m is formed of a slide whose dimensions are 5 cm x 5 cm.
  How far from the lens must the screen be placed?

Answer: 3.15 m
  ES Difficult

23.128: How far from a 50-mm focal length lens, such as is used in many
  35-mm cameras, must an object be positioned if it is to form a
  real image magnified in size by a factor of three?
  a) 46.2 mm   b) 52.2 mm   c) 57.5 mm   d) 66.7 mm

Answer: (d)
  MC Difficult

23.129: How far from a lens of focal length 50 mm must the object be
  placed if it is to form a virtual image magnified in size by a
  factor of three?
  a) 33.3 mm   b) 42.2 mm   c) 48.0 mm   d) 54.4 mm

Answer: (a)
  MC Difficult

23.130: If you stand in front of a double-concave lens, exactly at its
  center of curvature, your image will appear where you are, but
  upside-down.

Answer: False
  TF Moderate

23.131: An object is placed at the origin. A converging lens of focal
length 10 mm is placed at x = 40 mm on the x axis. A second
converging lens of focal length 20 mm is placed at x = 90 mm.
Graphically determine the size and location of the final image.

Answer:

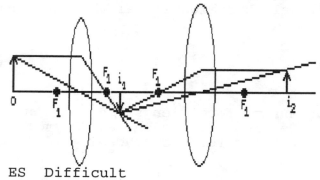

ES Difficult

23.132: A lamp is placed 1 m from a screen. Between the lamp and the
screen is placed a converging lens of focal length 24 cm. The
filament of the lamp can be imaged on the screen. As the lens
position is varied with respect to the lamp,
a) no sharp image will be seen for any lens position.
b) a sharp image will be seen when the lens is halfway between
the lamp and the screen.
c) a sharp image will be seen when the lens is 40 cm from the
lamp.
d) a sharp image will be seen when the lens is 60 cm from the
lamp.
e) a sharp image will be seen when the lens is either 40 cm from
the lamp or 60 cm from the lamp, but not otherwise.

Answer: (e)
MC Moderate

23.133: A diverging lens (f = -4 cm) is positioned 2 cm to the left of a
converging lens f = +6 cm). A 1-mm diameter beam of parallel
light rays is incident on the diverging lens from the left.
After leaving the converging lens, the outgoing rays
a) converge.
b) diverge.
c) form a parallel beam of diameter D > 1 mm.
d) form a parallel beam of diameter D < 1 mm.
e) will travel back toward the light source.

Answer: (c)
MC Difficult

23.134: Is it possible to see a virtual image?
a) No, since the rays that seem to emanate from a virtual image do not in fact emanate from the image.
b) No, since virtual images do not really exist.
c) Yes, the rays that appear to emanate from a virtual image can be focused on the retina just like those from an illuminated object.
d) Yes, since almost everything we see is virtual because most things do not themselves give off light, but only reflect light coming from some other source.
e) Yes, but only indirectly in the sense that if the virtual image is formed on a sheet of photographic film, one could later look at the picture formed.

Answer: (c)
MC   Moderate

23.135: Two thin double-convex (convex-convex) lenses are placed in contact. If each has a focal length of 20 cm, how would you expect the combination to function?
a) About like a single lens of focal length 20 cm
b) About like a single lens of focal length 40 cm
c) About like a single lens of focal length slightly greater than 20 cm
d) About like a single lens of focal length less than 20 cm

Answer: (d)
MC   Moderate

23.136: In a single-lens reflex camera the lens-film distance may be varied by sliding the lens forward or backward with respect to the camera housing. If, with such a camera, a fuzzy picture is obtained, this means that
a) the lens was too far from the film.
b) the lens was too close to the film.
c) too much light was incident on the film.
d) too little light was incident on the film.
e) one cannot say which of the above reasons is valid.

Answer: (e)
MC   Moderate

23.137: Two very thin lenses, each with focal length 20 cm, are placed in contact. What is the focal length of this compound lens?
a) 40 cm   b) 20 cm   c) 15 cm   d) 10 cm

Answer: (d)
MC   Moderate

23.138: Two thin lenses, of focal lengths f_1 and f_2 placed in contact with each other are equivalent to a single lens of focal length
    a) $f_1 + f_2$ b) $1/(f_1 + f_2)$
    c) $(f_1 + f_2)/f_1f_2$ d) $f_1f_2/(f_1 + f_2)$

Answer: (d)
    MC   Moderate

23.139: When an object is placed 60 cm from a converging lens, it forms a real image. When the object is moved to 40 cm from the lens, the image moves 10 cm farther from the lens. What is the focal length of the lens?
    a) 20 cm   b) 30 cm   c) 40 cm   d) 42 cm

Answer: (a)
    MC   Difficult

23.140: A converging lens with the same curvature on both sides and focal length 25 cm is to be made from crown glass (n = 1.52). What radius of curvature is required for each face?

Answer: 26 cm
    ES   Moderate

23.141: A double convex lens has faces of radii 18 cm and 20 cm. When an object is placed 24 cm from the lens, a real image is formed 32 cm from the lens. Determine (a) the focal length of the lens (b) the index of refraction of the lens material.

Answer: (a) 13.7 cm   (b) n = 1.69
    ES   Moderate

23.142: A double convex (convex-convex) thin lens has radii of curvature 46 cm, and is made of glass of index of refraction n = 1.60. What is the focal length?
    a) infinite   b) 38 cm   c) 30 cm   d) 18 cm

Answer: (b)
    MC   Easy

23.143: A plano-convex lens is to have a focal length of 40 cm. It is made of glass of index of refraction 1.65. What radius of curvature is required?
    a) 13 cm   b) 26 cm   c) 32 cm   d) 36 cm

Answer: (b)
    MC   Moderate

23.144: A +20 diopter lens has a different power in water.

Answer: True
    TF   Moderate

24. The Wave Nature of Light

24.1: A ray of light, which is traveling in a vacuum, is incident on a glass plate ($n = n_g$). For decreasing angles of incidence, the angle of refraction
a) increases, approaching the limiting value of 90°.
b) decreases, approaching the limiting value of $\sin^{-1}(1/n_g)$ degrees.
c) decreases, approaching the limiting value of $\sin^{-1}(n_g)$ degrees.
d) decreases, approaching the limiting value of zero degrees.

Answer: (d)
MC   Easy

24.2: For a thin spherical lens, the focal length of red light compared to the focal length of blue light is
a) greater.
b) smaller.
c) the same.
d) either greater, smaller, or the same. It depends on the type of glass from which the lens is made.

Answer: (a)
MC   Moderate

24.3: The wave theory of light is attributed to
a) Christian Huygens.   b) Isaac Newton.
c) Max Planck.   d) Albert Einstein.

Answer: (a)
MC   Easy

24.4: The particle theory of light is attributed to
a) Christian Huygens.   b) Isaac Newton.
c) Max Planck.   d) Albert Einstein.

Answer: (b)
MC   Easy

24.5: What principle is responsible for light spreading as it passes through a narrow slit?
a) refraction   b) polarization
c) diffraction   d) interference

Answer: (c)
MC   Easy

24.6: What principle is responsible for alternating light and dark bands when light passes through two or more narrow slits?
a) refraction
b) polarization
c) diffraction
d) interference

Answer: (d)
MC Easy

24.7: What principle is responsible for the fact that certain sunglasses can reduce glare from reflected surfaces?
a) refraction
b) polarization
c) diffraction
d) total internal reflection

Answer: (b)
MC Easy

24.8: The principle on which fiber optics is based is
a) refraction
b) polarization
c) dispersion
d) total internal reflection

Answer: (d)
MC Easy

24.9: The principle which allows a rainbow to form is
a) refraction
b) polarization
c) dispersion
d) total internal reflection

Answer: (c)
MC Easy

24.10: The principle on which lenses work is
a) refraction
b) polarization
c) dispersion
d) total internal reflection

Answer: (a)
MC Easy

24.11: The principle which explains why a prism separates white light into different colors is
a) refraction
b) polarization
c) dispersion
d) total internal reflection

Answer: (c)
MC Easy

24.12: Objects under water appear closer than they are because of refraction.

Answer: True
TF Easy

24.13: Fiber optics work on the principle of dispersion.

Answer: False
TF   Easy

24.14: Light forms light and dark bands when passing through multiple narrow slits because of diffraction.

Answer: False
TF   Easy

24.15: Rainbows form because of the principle of dispersion.

Answer: True
TF   Easy

24.16: Some sunglasses reduce light reflected off of surfaces because of polarization.

Answer: True
TF   Easy

24.17: Light has wavelength 600 nm in a vacuum. It passes into glass, which has an index of refraction of 1.50. What is the frequency of the light inside the glass?

Answer: 5.0×10^{14} Hz
ES   Easy

24.18: Light has a wavelength of 600 nm in a vacuum. It passes into glass, which has an index of refraction of 1.50. What is the speed of the light in the glass?

Answer: 2.0×10^8 m/s
ES   Easy

24.19: Light has a wavelength of 600 nm in a vacuum. It passes into glass, which has an index of refraction of 1.50. What is the wavelength of the light in the glass?

Answer: 400 nm
ES   Easy

24.20: For a certain type of glass, the index of refraction for red light is 1.522, and for blue light it is 1.531. A convex lens is made of this glass. For which color will the lens have a longer focal length?
   a) red
   b) blue
   c) Red and blue will have the same focal length.
   d) Impossible to determine from the information given.

Answer: (a)
   MC   Moderate

24.21: You are wearing polarized sunglasses while on a boat on a still lake. The glasses will absorb most of the light reflected off the lake.

Answer: True
   TF   Moderate

24.22: You are wearing polarized sunglasses while on a boat on a still lake. The glasses will absorb most of the direct sunlight when the sun is setting.

Answer: False
   TF   Easy

24.23: You are wearing polarized sunglasses while on a boat on a still lake. All of the light passing through glasses is polarized.

Answer: True
   TF   Easy

24.24: You are wearing polarized sunglasses while on a boat on a still lake. If you tilt your head, the amount of glare off the surface of the lake will decrease.

Answer: False
   TF   Easy

24.25: 400 nm of light falls on a single slit of width 0.1 mm. What is the angular width of the central diffraction peak?

Answer: 0.46°
   ES   Moderate

24.26: Two thin slits are 0.005 mm apart. Monochromatic light falls on these slits, and produces a fourth order interference fringe at an angle of 23.2°. What is the wavelength of the light?

Answer: 492 nm
   ES   Easy

24.27: Two thin slits are 0.05 mm apart. Monochromatic light of wavelength 634 nm falls on the slits. If there is a screen 6 m away, how far apart are adjacent interference fringes?

Answer: 7.6 cm
ES Moderate

24.28: A diffraction grating will only work if the wavelength of the light is larger than the distance between adjacent lines on the grating.

Answer: False
TF Moderate

24.29: A diffraction grating has 4000 lines per cm. The angle between the central maximum and the third order maximum is 36°. What is the wavelength of the light?

Answer: 490 nm
ES Moderate

24.30: Which of the following is a false statement?
a) All points on a given wave front have the same phase.
b) Rays are always perpendicular to wave fronts.
c) All wave fronts have the same amplitude.
d) The spacing between adjacent wave fronts is one-half wavelength.

Answer: (c)
MC Easy

24.31: Light of wavelength 550 nm in air is found to travel at 1.96×10^8 m/s in a certain liquid. Determine (a) the index of refraction of the liquid. (b) the frequency of the light in air. (c) the frequency of the light in the liquid. (d) the wavelength of the light in the liquid.

Answer: (a) 1.53   (b) 5.45×10^{14} Hz   (c) 5.45×10^{14} Hz   (d) 359 nm
ES Moderate

24.32: When a light wave enters into a medium of different optical density,
a) its speed and frequency change.
b) its speed and wavelength change.
c) its frequency and wavelength change.
d) its speed, frequency, and wavelength change.

Answer: (b)
MC Easy

24.33: Optical density is directly proportional to mass density.

Answer: False
　　　　TF　Easy

24.34: Laser light is invisible in air.

Answer: True
　　　　TF　Easy

24.35: When a beam of light (wavelength = 590 nm), originally traveling in air, enters a piece of glass (index of refraction 1.50), its frequency
　　　　a) increases by a factor of 1.50.
　　　　b) is reduced to 2/3 its original value.
　　　　c) is unaffected.
　　　　d) none of the above.

Answer: (c)
　　　　MC　Easy

24.36: When a beam of light (wavelength = 590 nm), originally traveling in air, enters a piece of glass (index of refraction 1.50), its wavelength
　　　　a) increases by a factor of 1.50.
　　　　b) is reduced to 2/3 its original value.
　　　　c) is unaffected.
　　　　d) none of the above.

Answer: (b)
　　　　MC　Easy

24.37: The index of refraction of diamond is 2.42. This means that a given frequency of light travels
　　　　a) 2.42 times faster in air than it does in diamond.
　　　　b) 2.42 times faster in diamond than it does in air.
　　　　c) 2.42 times faster in vacuum than it does in diamond.
　　　　d) 2.42 times faster in diamond than it does in vacuum.

Answer: (c)
　　　　MC　Easy

24.38: A beam of light ($f = 5 \times 10^{14}$ Hz) enters a piece of glass ($n = 1.5$). What is the frequency of the light while it is in the glass?
　　　　a) 5×10^{14} Hz　　　　　　　　b) 7.5×10^{14} Hz
　　　　c) 3.33×10^{14} Hz　　　　　　d) None of the above

Answer: (a)
　　　　MC　Easy

24.39: When a ray of light passes obliquely from one medium to another, which of the following changes?
    a) Direction of travel   b) Wavelength
    c) Speed   d) All of the above

Answer: (d)
    MC   Easy

24.40: Which color of light undergoes the greatest refraction when passing from air to glass?
    a) Red   b) Yellow   c) Green   d) Blue

Answer: (d)
    MC   Moderate

24.41: White light is monochromatic light.

Answer: False
    TF   Easy

24.42: A beam of white light is incident on a thick glass plate with parallel sides, at an angle between 0° and 90° with the normal. Which color emerges from the other side first?
    a) Red
    b) Green
    c) Violet
    d) None of the above. All colors emerge at the same time.

Answer: (d)
    MC   Difficult

24.43: When white light goes from air to glass, the green color component refracts more than the yellow color component.

Answer: True
    TF   Moderate

24.44: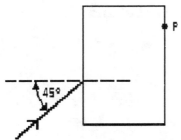

A light beam composed of red and blue light is incident upon a rectangular glass plate, as shown. The light rays that emerge into the air
a) are parallel, with the red beam displaced below the blue beam.
b) are parallel, with the blue beam displaced below the red beam.
c) are not parallel, with the blue beam displaced below the red beam.
d) are not parallel, with the red beam displaced below the blue beam.
e) none of the above is correct.

Answer: (b)
MC   Moderate

24.45: White light is
a) light of wavelength 550 nm, in the middle of the visible spectrum.
b) a mixture of all frequencies.
c) a mixture of red, green, and blue light.
d) the term used to describe very bright light.
e) the opposite (or complementary color) of black light.

Answer: (b)
MC   Easy

24.46: A parallel light beam containing two wavelengths, 480 nm and 700 nm, strikes a plain piece of glass at an angle of incidence of 60°. The index of refraction of the glass is 1.4830 at 480 nm and 1.4760 at 700 nm. Determine the angle between the two beams in the glass.

Answer: 0.196°
ES   Moderate

24.47: A ray of light consisting of blue light (wavelength 480 nm) and red light (wavelength 670 nm) is incident on a thick piece of glass at 80°. What is the angular separation between the refracted red and refracted blue beams while they are in the glass? (The respective indices of refraction for the blue light and the red light are 1.4636 and 1.4561.)
a) 0.269°   b) 0.330°   c) 0.341°   d) 0.455°

Answer: (a)
MC   Moderate

24.48: You can only see a rainbow if the sun is behind you.

Answer: True
TF   Easy

24.49: The higher the sun is up in the sky, the more the rainbow can be seen from the ground.

Answer: False
TF   Moderate

24.50: The color on the outer edge of the primary rainbow is red.

Answer: True
TF   Moderate

24.51: Two light sources are said to be coherent if they
a) are of the same frequency.
b) are of the same frequency, and maintain a constant phase difference.
c) are of the same amplitude, and maintain a constant phase difference.
d) are of the same frequency and amplitude.

Answer: (b)
MC   Easy

24.52: Two beams of coherent light travel different paths arriving at point P. If the maximum constructive interference is to occur at point P, the two beams must
a) arrive 180° out of phase.
b) arrive 90° out of phase.
c) travel paths that differ by a whole number of wavelengths.
d) travel paths that differ by an odd number of half-wavelengths.

Answer: (c)
MC   Easy

24.53: One beam of coherent light travels path P_1 in arriving at point Q and another coherent beam travels path P_2 in arriving at the same point. If these two beams are to interfere destructively, the path difference $P_1 - P_2$ must be equal to
a) an odd number of half-wavelengths.
b) zero.
c) a whole number of wavelengths.
d) a whole number of half-wavelengths.

Answer: (a)
MC   Easy

24.54: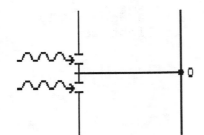

Monochromatic light from a distant source is incident on two parallel narrow slits. After passing through the slits the light strikes a screen, as shown in the sketch. What will be the nature of the pattern of light observed on the screen?
a) The screen will be most brightly illuminated at point O, with the intensity decreasing slowly and uniformly as you move outward from point O.
b) A series of alternating light and dark bands.
c) A rainbow of colored lines will be seen spreading out on either side of point O.
d) Two bright bands of light, one in line with each slit.
e) The screen will be uniformly illuminated except for two dark bands, one in line with each slit.

Answer: (b)
MC   Easy

24.55: In a double-slit experiment, it is observed that the distance between adjacent maxima on a remote screen is 1 cm. What happens to the distance between adjacent maxima when the slit separation is cut in half?
a) It increases to 2 cm.   b) It increases to 4 cm.
c) It decreases to 0.5 cm.   d) It decreases to 0.25 cm.

Answer: (a)
MC   Easy

24.56: At the first maxima on either side of the central bright spot in a double-slit experiment, light from each opening arrives
a) in phase.   b) 90° out of phase.
c) 180° out of phase.   d) none of the above.

Answer: (a)
MC   Easy

24.57: At the second maxima on either side of the central bright spot in a double-slit experiment, light from
a) each opening travels the same distance.
b) one opening travels twice as far as light from the other opening.
c) one opening travels one wavelength of light farther than light from the other opening.
d) one opening travels two wavelengths of light farther than light from the other opening.

Answer: (d)
MC   Moderate

24.58: The separation between adjacent maxima in a double-slit interference pattern using monochromatic light is
a) greatest for red light.
b) greatest for green light.
c) greatest for blue light.
d) the same for all colors of light.

Answer: (a)
MC   Easy

24.59: In a diffraction experiment, light of 600 nm wavelength produces a first-order maximum 0.35 mm from the central maximum on a distant screen. A second monochromatic source produces a third-order maximum 0.87 mm from the central maximum when it passes through the same diffraction grating. What is the wavelength of the light from the second source?

Answer: 497 nm
ES   Moderate

24.60: In a double-slit experiment, the slit separation is 2 mm, and two wavelengths, 750 nm and 900 nm, illuminate the slits. A screen is placed 2 m from the slits. At what distance from the central maximum on the screen will a bright fringe from one pattern first coincide with a bright fringe from the other?

Answer: 4.5 mm
ES   Moderate

24.61: A single slit, which is 0.05 mm wide, is illuminated by light of 550 nm wavelength. What is the angular separation between the first two minima on either side of the central maximum?

Answer: 0.630°
ES   Moderate

24.62: If two light waves are coherent, which of the following is NOT necessary?
a) They must have the same frequency.
b) They must have the same wavelength.
c) They must have the same amplitude.
d) They must have the same velocity.
e) They must have a constant phase difference at every point in space.

Answer: (c)
MC   Easy

24.63: Why would it be impossible to obtain interference fringes in a
double-slit experiment if the separation of the slits is less
than the wavelength of the light used?
a) The very narrow slits required would generate many different
wavelengths, thereby washing out the interference pattern.
b) The two slits would not emit coherent light.
c) The fringes would be too close together.
d) In no direction could a path difference as large as one
wavelength be obtained, and this is needed if a bright fringe,
in addition to the central fringe, is to be observed.

Answer: (d)
MC   Easy

24.64: What do we mean when we say that two light rays striking a screen
are in phase with each other?
a) When the electric field due to one is a maximum, the electric
field due to the other is also a maximum, and this relation is
maintained as time passes.
b) They are traveling at the same speed.
c) They have the same wavelength.
d) They alternately reinforce and cancel each other.

Answer: (a)
MC   Easy

24.65: The colors on an oil slick are caused by reflection and
a) diffraction.   b) interference.
c) refraction.   d) polarization.

Answer: (b)
MC   Easy

24.66: When a beam of light, which is traveling in glass, strikes an air
boundary, there is
a) a 90° phase change in the reflected beam.
b) no phase change in the reflected beam.
c) a 180° phase change in the reflected beam.
d) a 45° phase change in the reflected beam.

Answer: (b)
MC   Easy

24.67: When a beam of light, which is traveling in air, is reflected by
a glass surface, there is
a) a 90° phase change in the reflected beam.
b) no phase change in the reflected beam.
c) a 180° phase change in the reflected beam.
d) a 45° phase change in the reflected beam.

Answer: (c)
MC   Easy

24.68: A soap film is being viewed in white light. As the film becomes very much thinner than the wavelength of blue light, the film
a) appears black because it reflects no visible light.
b) appears white because it reflects all wavelengths of visible light.
c) appears blue since all other colors are transmitted.
d) appears red since all other colors are transmitted.

Answer: (a)
    MC Easy

24.69: In terms of the wavelength of light in soapy water, what is the minimum thickness of soap film that will reflect a given wavelength of light?
a) One-fourth wavelength
b) One-half wavelength
c) One wavelength
d) There is no minimum thickness.

Answer: (a)
    MC Moderate

24.70: In terms of the wavelength of light in magnesium fluoride, what is the minimum thickness of magnesium fluoride coating that must be applied to a glass lens to make it non-reflecting for that wavelength? (The index of refraction of magnesium fluoride is intermediate to that of glass and air.)
a) One-fourth wavelength
b) One-half wavelength
c) One wavelength
d) There is no minimum thickness.

Answer: (a)
    MC Moderate

24.71: A convex lens is placed on a flat glass plate and illuminated from above with monochromatic red light. When viewed from above, concentric bands of red and dark are observed. What does one observe at the exact center of the lens where the lens and the glass plate are in direct contact?
a) A bright red spot
b) A dark spot
c) A rainbow of color
d) A bright spot that is some color other than red

Answer: (b)
    MC Moderate

24.72: We have seen that two monochromatic light waves can interfere constructively or destructively, depending on their phase difference. One consequence of this phenomena is
a) the colors you see when white light is reflected from a soap bubble.
b) the appearance of a mirage in the desert.
c) a rainbow.
d) the way in which polaroid sunglasses work.
e) the formation of an image by a converging lens, such as the lens in your eye.

Answer: (a)
MC   Easy

24.73: A soap bubble has an index of refraction of 1.33. What minimum thickness of this bubble will ensure maximum reflectance of normally incident 530 nm wavelength light?

Answer: 99.6 nm
ES   Moderate

24.74: A lens is coated with material of index of refraction 1.38. What thickness of coating should be used to give maximum transmission at a wavelength of 530 nm?

Answer: 96 nm
ES   Moderate

24.75: Light of wavelength 500 nm illuminates a soap film (n = 1.33). What is the minimum thickness of film that will give an interference when the light is incident normally on it?

Answer: 94 nm
ES   Moderate

24.76: A glass plate 2.5 cm long is separated from another glass plate at one end by a strand of someone's hair (diameter 0.001 cm). How far apart are the adjacent interference bands when viewed with light of wavelength 600 nm?

Answer: 0.75 mm
ES   Moderate

24.77: After a rain, one sometimes sees brightly colored oil slicks on the road. These are due to
a) interference effects.
b) polarization effects.
c) diffraction effects.
d) selective absorption of different wavelengths by oil.
e) none of the above.

Answer: (a)
MC   Easy

24.78: Consider two diffraction gratings; one has 4000 lines per cm and the other one has 6000 lines per cm.
a) The 4000-line grating produces the greater dispersion.
b) The 6000-line grating produces the greater dispersion.
c) Both gratings produce the same dispersion, but the orders are sharper for the 4000-line grating.
d) Both gratings produce the same dispersion, but the orders are sharper for the 6000-line grating.

Answer: (b)
MC   Easy

24.79: Consider two diffraction gratings with the same slit separation, the only difference being that one grating has 3 slits and the other 4 slits. If both gratings are illuminated with a beam of the same monochromatic light,
a) the grating with 3 slits produces the greater separation between orders.
b) the grating with 4 slits produces the greater separation between orders.
c) both gratings produce the same separation between orders.
d) both gratings produce the same separation between orders, but the orders are better defined with the 4-slit grating.

Answer: (d)
MC   Moderate

24.80: A diffraction grating has 6000 lines per centimeter ruled on it. What is the angular separation between the second and the third orders on the same side of the central order when the grating is illuminated with a beam of light of wavelength 550 nm?

Answer: 40.5°
ES   Moderate

24.81: When light illuminates a grating with 7000 lines per centimeter, its second maximum is at 62.4°. What is the wavelength of the light?

Answer: 633 nm
ES   Easy

24.82: An ideal polarizer is placed in a beam of unpolarized light and the intensity of the transmitted light is 1. A second ideal polarizer is placed in the beam with its referred direction rotated 40° to that of the first polarizer. What is the intensity of the beam after it has passed through both polarizers?
a) 0.766   b) 0.643   c) 0.587   d) 0.413

Answer: (c)
MC   Easy

24.83: For a beam of light, the direction of polarization is defined as
    a) the beam's direction of travel.
    b) the direction of the electric field's vibration.
    c) the direction of the magnetic field's vibration.
    d) the direction that is mutually perpendicular to the electric and magnetic field vectors.

Answer: (b)
    MC Easy

24.84: A polarizer (with its preferred direction rotated 30° to the vertical) is placed in a beam of unpolarized light of intensity 1. After passing through the polarizer, the beam's intensity is
    a) 0.250   b) 0.500   c) 0.866   d) 0.750

Answer: (b)
    MC Moderate

24.85: What is the Brewster's angle for light traveling in vacuum and reflecting off a piece of glass of index of refraction 1.52?
    a) 48.9°   b) 41.1°   c) 33.3°   d) 56.7°

Answer: (d)
    MC Easy

24.86: Two emerging beams of light produced by a birefringent crystal
    a) are at different frequencies.
    b) will produce an interference pattern when recombined.
    c) are polarized in the same direction.
    d) are polarized in mutually perpendicular directions.

Answer: (d)
    MC Easy

24.87: Suppose that you take the lenses out of a pair of polarized sunglasses and place one on top of the other. Rotate one lens 90° with respect to the normal position of the other lens. Early in the morning, look directly overhead at the sunlight coming down. What would you see?
    a) The lenses would look completely dark, since they would transmit no light.
    b) You would see some light, and the brightness would be a function of how the lenses were oriented with respect to the incoming light.
    c) You would see light with intensity reduced to about 50% of what it would be with one lens.
    d) You would see some light, and the intensity would be the same as if you had looked through only one lens.
    e) You would see some light, and the intensity would be greater than if you looked through only one lens.

Answer: (a)
    MC Easy

24.88: A beam of light passes through a polarizer and then an analyzer. In this process, the intensity of the light transmitted is reduced to 10% of the intensity incident on the analyzer. What is the angle between the axes of the polarizer and the analyzer?

Answer: 71.6°
ES   Moderate

24.89: Sunlight reflected from the surface of a lake
a) is unpolarized.
b) tends to be polarized with its electric field vector parallel to the surface of the lake.
c) tends to be polarized with its electric field vector perpendicular to the surface of the lake.
d) has undergone refraction by the surface of the lake.
e) none of the above is true.

Answer: (b)
MC   Easy

24.90: The polarization of sunlight is greatest at
a) sunrise.   b) sunset.
c) both sunrise and sunset.   d) midday.

Answer: (c)
MC   Moderate

24.91: If sunlight of color B is scattered through an angle 16 times greater than sunlight of color A, then the wavelength of color B is
a) twice that of color A.   b) 1/2 that of color A.
c) 16 times that of color A.   d) 1/16 that of color A.
e) none of the above.

Answer: (b)
MC   Moderate

24.92: On a clear day, the sky appears to be more blue toward the zenith (overhead) than it does toward the horizon. This occurs because
a) the atmosphere is denser higher up than it is at the earth's surface.
b) the temperature of the upper atmosphere is higher than it is at the earth's surface.
c) the sunlight travels over a longer path at the horizon, resulting in more absorption.
d) none of the above is true.

Answer: (c)
MC   Moderate

24.93: In single-slit diffraction
   a) the intensities of successive maxima are roughly the same, falling off only gradually as one goes away from the central maximum.
   b) the central maximum is about as wide as the other maxima.
   c) the central maximum is about twice as wide as the other maxima.
   d) the central maximum is much narrower than the other maxima.
   e) the slit width must be less than one wavelength for a diffraction pattern to be apparent.

Answer: (c)
   MC   Moderate

25. Optical Instruments

25.1: A near-sighted person has a far point of 18 cm. What lens (in diopters) will allow this person to see distant objects clearly?
   a) +5.56 dipoters
   b) -5.56 diopters
   c) +0.056 diopters
   d) -0.056 diopters

Answer: (b)
MC   Moderate

25.2: A farsighted person can read a newspaper held 25 cm from his eyes, if he wears glasses of +6 diopters. What is this person's near point?
   a) 4.17 cm
   b) 25 cm
   c) 31 cm
   d) 150 cm

Answer: (d)
MC   Moderate

25.3: A magnifying lens has a focal length of 10 cm. A person has a near point of 25 cm. What is the magnification of the lens for that person when their eyes are focused at infinity?

Answer: 2.5
ES   Moderate

25.4: A magnifying lens has a focal length of 10 cm. A person has a near point of 25 cm. What is the magnification of the lens for that person when their eyes are focused at their near point?

Answer: 3.5
ES   Moderate

25.5: A microscope has an objective lens of focal length 1.4 mm and an eyepiece of focal length 20 mm, adjusted for minimum eyestrain. A blood sample is placed 1.5 mm from the objective.
   a) How far apart are the lenses?
   b) What is the overall magnification?

Answer: a)   41 mm
        b)   175
ES   Moderate

25.6: You have available lenses of focal lengths 2 cm, 4 cm, 8 cm, and 16 cm. If you were to use any two of these lenses to build a telescope,
a) what is the maximum magnification you could achieve?
b) What is the lens separation for the maximum magnification telescope?

Answer: a)   8
        b)   18 cm
ES   Easy

25.7: Spherical lenses suffer from
   a) both spherical and chromatic aberration.
   b) spherical aberration, but not chromatic aberration.
   c) chromatic aberration, but not spherical aberration.
   d) neither spherical nor chromatic aberration.

Answer: (a)
   MC   Easy

25.8: Spherical aberration can occur in mirrors as well as lenses.

Answer: True
   TF   Easy

25.9: Spherical mirrors suffer from
   a) both spherical and chromatic aberration.
   b) spherical aberration, but not chromatic aberration.
   c) chromatic aberration, but not spherical aberration.
   d) neither spherical nor chromatic aberration.

Answer: (b)
   MC   Easy

25.10: Chromatic aberration occurs because of dispersion.

Answer: True
   TF   Easy

25.11:

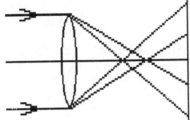

When two parallel white rays pass through the outer edges of a converging glass lens, chromatic aberration will cause colors to appear on the screen in the following order, from the top down:
   a) red, blue, red, blue.
   b) red, blue, blue, red.
   c) blue, red, blue, red.
   d) blue, red, red, blue.
   e) blue, blue, red, red.

Answer: (d)
   MC   Moderate

25.12: Compare two diverging lenses similar except that lens B is rated at 20 diopters, whereas lens A is rated at 10 diopters. The focal length of lens B is
   a) one-fourth of the focal length of lens A.
   b) one-half of the focal length of lens A.
   c) twice the focal length of lens A.
   d) four times the focal length of lens A.

Answer: (b)
     MC   Easy

25.1: **MATCH THE EYE PART'S FUNCTION TO ITS NAME.**

25.13: iris

Answer: a circular diaphragm
     MA   Easy

25.14: pupil

Answer: the central hole
     MA   Easy

25.15: retina

Answer: a light-sensitive surface, upon which the image falls
     MA   Easy

25.16: cornea

Answer: a curved transparent tissue where light first enters the eye
     MA   Easy

25.17: lens

Answer: adjusted by ciliary muscles
     MA   Easy

25.18: The aperture in a camera plays the same role as the retina in the eye.

Answer: False
     TF   Easy

25.19: 30/40 vision means that one eye can see clearly at 30 ft, and the other eye can see clearly at 40 ft.

Answer: False
     TF   Easy

25.20: What power lens is needed to correct for nearsightedness where the uncorrected far point is 75 cm?
   a) -0.75 D
   b) +1.33 D
   c) -1.33 D
   d) None of the above

Answer: (c)
   MC   Moderate

25.21: What power lens is needed to correct for farsightedness where the uncorrected near point is 75 cm?
   a) -2.67 D
   b) +2.67 D
   c) +5.33 D
   d) None of the above

Answer: (b)
   MC   Moderate

25.22: A nearsighted man wears -3.0 D lenses. With these lenses, his corrected near point is 25 cm. What is his uncorrected near point?

Answer: 14.3 cm
   ES   Moderate

25.23: A nearsighted person has a near point of 12 cm and a far point of 17 cm. If the lens is 2 cm from the eye, (a) what lens power will enable this person to see distant objects clearly? (b) what then will be the new near point?

Answer: (a) -6.7 D   (b) 30 cm in front of the lens
   ES   Difficult

25.24: A person uses corrective glasses of power -8.5 D. (a) Is this person nearsighted or farsighted? (b) If the glasses are worn 2 cm from the eye, what is the person's far point without glasses?

Answer: (a) nearsighted   (b) 14 cm
   ES   Difficult

25.25: A fish's eye is well-suited to seeing under water, but what can you say about his vision if he is taken out of water?
   a) His acuity (sharpness of vision) would be greater.
   b) He would suffer from astigmatism.
   c) He would be nearsighted.
   d) He would be farsighted.
   e) His vision would be limited to a cone of half-angle of about 49°.

Answer: (c)
   MC   Moderate

25.26: The principal refraction of light by the eye occurs at the
a) cornea.   b) lens.   c) retina.   d) iris.

Answer: (a)
MC   Easy

25.27: Rays that pass through a lens very close to the lens axis are
more sharply focused than those that are very far from the axis.
The inability of a lens to sharply focus off-axis rays is called
spherical aberration. This effect helps us to understand why
a) we become more farsighted as we grow older.
b) we become more nearsighted as we grow older.
c) we can see only in black and white in dim light.
d) it is easier to read in bright light than in dim light.
e) a moving object is more readily detected than a stationary
one.

Answer: (d)
MC   Moderate

25.28: If a person's eyeball is too short from front to back, the person
is likely to suffer from
a) astigmatism.   b) spherical aberration.
c) farsightedness.   d) nearsightedness.

Answer: (c)
MC   Easy

25.29: If a person's eyeball is too long from front to back, the person
is likely to suffer from
a) spherical aberration.   b) nearsightedness.
c) farsightedness.   d) astigmatism.

Answer: (b)
MC   Easy

25.30: Nearsightedness can usually be corrected with
a) converging lenses.   b) diverging lenses.
c) achromatic lenses.   d) cylindrical lenses.

Answer: (b)
MC   Easy

25.31: The near point of a farsighted person is 100 cm. She places
reading glasses close to her eyes, and with them she can
comfortably read a newspaper at a distance of 25 cm. What lens
power is required?
a) +2.5 D   b) +3.0 D   c) +3.2 D   d) -2.0 D

Answer: (b)
MC   Moderate

25.32: Farsightedness can usually be corrected with
    a) cylindrical lenses.   b) achromatic lenses.
    c) diverging lenses.   d) converging lenses.

Answer: (d)
    MC   Easy

25.33: If the human eyeball is too short from front to back, this gives
    rise to a vision defect that can be corrected by using
    a) convex-convex eyeglasses.
    b) concave-convex eyeglasses.
    c) cylindrical eyeglasses.
    d) contact lenses, but no ordinary lenses.
    e) shaded glasses (i.e., something that will cause the iris to
    dilate more).

Answer: (a)
    MC Moderate

25.34: In which of the following ways is a camera different from the
    human eye?
    a) The camera always forms an inverted image, the eye does not.
    b) The camera always forms a real image, the eye does not.
    c) The camera utilizes a fixed focal length lens, the eye does
    not.
    d) For the camera, the image magnification is greater than one,
    but for the eye the magnification is less than one.
    e) A camera cannot focus on objects at infinity but the eye can.

Answer: (c)
    MC Moderate

25.35: What is the maximum angular magnification of a magnifying glass
    of focal length 10 cm? (Assume the near point is at 25 cm.)

Answer: 3.5
    ES Moderate

25.36: A magnifying glass of focal length 15 cm is used to examine an
    old manuscript. (a) What is the maximum magnification given by
    the lens? (b) What is the magnification for relaxed eye-viewing
    (image at infinity)?

Answer: (a) 2.67   (b) 1.67
    ES Moderate

25.37: The first important discoveries made with a telescope (in 1609)
    were done by
    a) Hale.   b) Newton.
    c) Cassegrain.   d) Galileo.

Answer: (d)
    MC Easy

25.38: The objective of a telescope has a focal length of 100 cm and its eyepiece has a focal length of 5 cm. What is the magnification of this telescope when viewing an object at infinity?
a) 20
b) 0.05
c) 500
d) None of the above

Answer: (a)
MC   Easy

25.39: A refracting telescope has an eyepiece of focal length -8 cm and an objective of focal length 36 cm. What is the magnification?

Answer: 4.5
ES   Easy

25.40: A student wishes to build a telescope. She has available an eyepiece of focal length of -5 cm. What focal length objective is needed to obtain a magnification of 10 x?

Answer: 50 cm
ES   Easy

25.41: An objective lens of a telescope has a 1.9-m focal length. When viewed through this telescope, the moon appears 5.25 times larger than normal. How far apart are the objective lens and the eyepiece when this instrument is focused on the moon?

Answer: 2.26 m
ES   Moderate

25.42: Consider the image formed by a refracting telescope. Suppose an opaque screen is placed in front of the lower half of the objective lens. What effect will this have?
a) The top half of the image will be blacked out.
b) The lower half of the image will be blacked out.
c) The entire image will be blacked out, since the entire lens is needed to form an image.
d) The image will appear as it would if the objective were not blocked, but it will be dimmer.
e) There will be no noticeable difference in the appearance of the image with the objective partially blocked or not.

Answer: (d)
MC   Moderate

25.43: If the diameter of a radar dish is doubled, what happens to its resolving power assuming that all other factors remain unchanged? Its resolving power
   a) quadruples.
   b) doubles.
   c) halves.
   d) is reduced to one-quarter its original value.

Answer: (b)
   MC   Moderate

25.44: What is the smallest angular separation (in minutes of arc) for which two point sources can be resolved using 3-cm microwaves and a 3-m diameter radar dish?

Answer: 41.9'
   ES   Moderate

25.45: In which of the following is diffraction NOT exhibited?
   a) Viewing a light source through a small pinhole
   b) Examining a crystal by X-rays
   c) Using a microscope under maximum magnification
   d) Resolving two nearby stars with a telescope
   e) Determining the direction of polarization with a birefringent crystal

Answer: (e)
   MC   Moderate

25.46: The world's largest refracting telescope is operated at the Yerkes Observatory in Wisconsin. It has an objective of diameter 102 cm. Suppose such an instrument could be mounted on a spy satellite at an elevation of 300 km above the surface of the earth. What is the minimum separation of two objects on the ground if their images are to be clearly resolved by this lens? (Assume an average wavelength of 550 nm for white light.)

Answer: 0.20 m
   ES   Moderate

25.47: With what color light would you expect to be able to see the greatest detail when using a microscope?
   a) red, because of its long wavelength
   b) red, because it is refracted less than other colors by glass
   c) blue, because of its shorter wavelength
   d) blue, because it is brighter
   e) The color makes no difference in the resolving power, since this is determined only by the diameter of the lenses.

Answer: (c)
   MC   Moderate

25.48: An important reason for using a very large diameter objective in an astronomical telescope is
a) to increase the magnification.
b) to increase the resolution.
c) to form a virtual image, which is easier to look at.
d) to increase the width of the field of view.
e) to increase the depth of the field of view.

Answer: (b)
MC   Easy

25.49: Radio waves are diffracted by large objects such as buildings, whereas light is not. Why is this?
a) Radio waves are unpolarized, whereas light is plane polarized.
b) The wavelength of light is much smaller than the wavelength of radio waves.
c) The wavelength of light is much greater than the wavelength of radio waves.
d) Radio waves are coherent, and light is usually not coherent.

Answer: (b)
MC   Easy

25.50: A person gazes at a very distant light source. If she now holds up two fingers, with a very small gap between them, and looks at the light source, she will see
a) the same thing as without the fingers, but dimmer.
b) a series of bright spots.
c) a sequence of closely spaced bright lines.
d) a hazy band of light varying from red at one side to blue or violet at the other.

Answer: (c)
MC   Easy

25.51: Railroad tracks are spaced about 1.7 m apart. On a clear day, how high could a plane fly before the pilot is no longer able to resolve the two separate rails? (Assume that the diameter of the pupil is about 3 mm and take as the wavelength that to which the eye is most sensitive, namely 530 nm.)

Answer: About 8 km
ES   Moderate

25.52: If the pupil diameter of the dark-adapted eye is 5 mm, what is the maximum distance at which two point sources of light that are 3 mm apart can still be resolved? Assume light of wavelength 530 nm. (You might try to confirm this by using a cardboard mask with two small pinholes placed in front of a light source that is completely enclosed.)

Answer: 23 m
ES   Moderate

25.53: Assuming the film used has uniform sensitivity throughout the visible spectrum, in which of the following cases would you be able to best distinguish between two closely spaced stars? (The lens referred to is the objective lens of the telescope used.)
   a) Use a large lens and blue light.
   b) Use a large lens and red light.
   c) Use a small lens and blue light.
   d) Use a small lens and red light.

Answer: (a)
   MC Moderate

25.54: The resolving power of a microscope refers to the ability to
   a) distinguish objects of different colors.
   b) form clear images of two points that are very close together.
   c) form a very large image.
   d) form a very bright image.

Answer: (b)
   MC Easy

25.55: If a red filter were placed in front of blue dungarees, you would see
   a) red.   b) blue.   c) green.   d) white.   e) black.

Answer: (e)
   MC Easy

25.56: If you wear green sunglasses, and look at equations written in white chalk on a chalkboard, you would see the chalk as
   a) green.   b) white.   c) black.   d) red.   e) yellow.

Answer: (a)
   MC Easy

26. Special Theory of Relativity

26.1: By what factor has the mass of an electron increased when it is accelerated to a speed of 0.99 c?

Answer: 7.1
ES   Easy

26.2: During a reaction, an element loses 4.8×10^{-28} kg of mass. How much energy (in Joules) is released?
   a) 4.32×10^{-11} J
   b) 1.44×10^{-19} J
   c) 1.6×10^{-36} J
   d) 5.3×10^{-45} J

Answer: (a)
MC   Easy

26.3: During a reaction, 1.7×10^{-4} J of energy is released. What change of mass would cause this?
   a) 5.1×10^{-4} kg
   b) 1.5×10^{-13} kg
   c) 4.8×10^{-18} kg
   d) 1.9×10^{-21} kg

Answer: (d)
MC   Easy

26.4: How much energy would be released if 2 kg of material was lost during a reaction?
   a) 1.8×10^{17} J
   b) 1.5×10^{16} J
   c) 6.0×10^{8} J
   d) 4.7×10^{-8} J

Answer: (a)
MC   Easy

26.5: What energy is released (in MeV) during a reaction in which 1.67×10^{-25} kg of material is totally converted to energy?
   a) 5.0×10^{-14} MeV
   b) 4.1×10^{-7} MeV
   c) 3.1×10^{-4} MeV
   d) 9.4×10^{4} MeV

Answer: (d)
MC   Moderate

26.6: How much mass is lost during a reaction in which 1.7×10^{8} MeV of energy is released?
   a) 1.8×10^{-8} kg
   b) 5.7×10^{-9} kg
   c) 1.9×10^{-17} kg
   d) 3.0×10^{-22} kg

Answer: (d)
MC   Moderate

26.7: A free neutron has a lifetime of 900 s. Suppose a neutron moves with a speed of 2.7×10^8 m/s relative to an observer. What will the observer measure the lifetime of the neutron to be?

Answer: 2060 s
ES   Moderate

26.8: A subatomic particle called a lambda has a rest mass of 1.852×10^{-27} kg. What is the rest energy of the lambda in MeV?

Answer: 1040 MeV
ES   Moderate

26.9: A muon is a subatomic particle with a rest energy of 106 MeV. If a muon is traveling at 0.98c, what is its total energy?

Answer: 533 MeV
ES   Moderate

26.10: A muon is a subatomic particle with a lifetime of 2.2×10^{-6} s. If a muon is moving with a speed of 0.98c relative to a stationary observer, what will the observer measure the muon's lifetime to be?

Answer: 11.1×10^{-6} s
ES   Moderate

26.11: A muon is a subatomic particle with a lifetime of 2.2×10^{-6} s. If a muon is traveling at a speed of 0.98c relative to a stationary observer, how far will the observer see the muon travel before it decays?

Answer: 3250 m
ES   Difficult

26.12: Moving clocks run slow because of the principle of relativity.

Answer: False
TF   Moderate

26.13: Observers on earth and on a rocket moving away from earth at high speed both get the same result when measuring the speed of sunlight because of the principle of relativity.

Answer: True
TF   Moderate

26.14: Identical experiments are performed on earth and on a rocket moving at high constant velocity. Both experiments will obtain the same results.

Answer: True
TF Moderate

26.15: The twin paradox is due to time dilation.

Answer: True
TF Easy

26.16: A high-speed rocket is shortened as measured by a stationary observer because of the equivalence of mass and energy.

Answer: False
TF Easy

26.17: A short-lived particle will travel farther before decaying when traveling at high speeds due to length contraction.

Answer: False
TF Moderate

26.18: At high speeds, kinetic energy is no longer given by $1/2\ mv^2$ because of the equivalence of mass and energy.

Answer: True
TF Moderate

26.19: You are in a rocket traveling away from the sun at 0.95c. You measure the speed of light from the sun to be
a) 0.05c
b) 0.95c
c) 1.00c
d) Cannot be determined without knowing the frequency of the light.

Answer: (c)
MC Easy

26.20: The speed of light is observed to be the same by all observers at rest or moving at constant velocity.

Answer: True
TF Easy

26.21: The laws of physics differ depending on whether you are at rest or moving at a constant velocity.

Answer: False
TF Moderate

26.22: The time interval between two events A and B is called the proper time if A and B occur at the same place.

Answer: True
TF Moderate

26.23: A set of twins, Andrea and Courtney, are initially 10 years old. While Courtney remains on Earth, Andrea rides on a space ship which travels away from earth at a speed of 0.6c for five years (as measured by Courtney), then turns around and comes back at 0.6c. When Andrea returns, Courtney is 20 years old. How old is Andrea upon her return?

Answer: 18 years
ES Difficult

26.24: An electron has a rest energy of 511 keV. It is traveling down a linear accelerator so fast that its energy is 40.88 MeV. The linear accelerator is 1000 m long. To a stationary observer, how long does it take the electron to travel the length of the linear accelerator?

Answer: 3.33×10^{-6} s
ES Moderate

26.25: An electron has a rest energy of 511 keV. It is traveling down a linear accelerator so fast that its energy is 40.88 MeV. The linear acclerator is 1000 m long. According to an observer riding on the electron, how long is the linear accelerator?

Answer: 1.25 m
ES Moderate

26.26: An electron has a rest mass of 511 keV. It is traveling down a linear accelerator so fast that its energy is 40.88 MeV. The linear accelerator is 1000 m long. According to an observer on the electron, how long does the trip down the accelerator take?

Answer: 4.17×10^{-8} s
ES Difficult

26.27: An electron has a rest mass of 511 keV. It is traveling down a linear accelerator so fast that its energy is 40.88 MeV. What is the speed of the electron?

Answer: 0.9999219c
ES Moderate

26.28: A space probe has a rest mass of 5000 kg. It travels away from the earth at a speed of 3,600,000 km/h. By how much does its mass increase, as measured from earth?

Answer: 27.8 g
ES   Moderate

26.29: How much energy is required to accelerate a 5000 kg space probe to a speed of 0.9c?

Answer: 5.8×10^{20} J
ES   Moderate

26.30: A boat can travel 4 m/s in still water. With what speed, relative to the shore, does it move in a river that is flowing at 1 m/s if the boat is heading (a) upstream. (b) downstream. (c) straight across the river.

Answer: (a) 3 m/s   (b) 5 m/s   (c) 4.12 m/s
ES   Easy

26.31: The Michelson-Morley experiment was designed to measure
a) the relativistic mass of the electron.
b) the relativistic energy of the electron.
c) the velocity of the earth relative to the ether.
d) the acceleration of gravity on the earth's surface.

Answer: (c)
MC   Easy

26.32: Michelson and Morley concluded from the results of their experiment that
a) the experiment was a failure since there was no detectable shift in the interference pattern.
b) the experiment was successful in not detecting a shift in the interference pattern.
c) the experiment was a failure since they detected a shift in the interference pattern.
d) the experiment was successful in detecting a shift in the interference pattern.

Answer: (a)
MC   Easy

26.33: You can build an interferometer yourself if you use the following components:
a) a light source, a detector screen, a partially silvered mirror, a flat mirror, and a glass plate.
b) a light source, a detector screen, two partially silvered mirrors, and a glass plate.
c) a light source, a detector screen, two partially silvered mirrors, a flat mirror, and a glass plate.
d) a light source, a detector screen, a partially silvered mirror, two flat mirrors, and a glass plate.

Answer: (d)
MC   Moderate

26.34: One consequence of Einstein's theory of special relativity is that absolute velocity of stars can be measured with respect to the ether wind.

Answer: False
TF   Easy

26.35: The gamma factor is defined as $\gamma \equiv 1/\sqrt{1 - (v/c)^2}$, therefore gamma ($\gamma$)
a) can be zero.
b) can be a negative number.
c) can be any number greater than or equal to zero.
d) can be any number greater than or equal to one.
e) cannot equal one.

Answer: (d)
MC   Easy

26.36: How fast would a rocket ship have to move to contract to half of its proper length (as observed by a stationary object)?

Answer: 0.87 c
ES   Moderate

26.37: A radar operator on earth sees two spaceships moving straight at each other, each with speed 0.6 c. With what speed does the pilot of one ship see the other ship approaching?

Answer: 0.88 c
ES   Easy

26.38: How fast should a moving clock travel if it is to be observed by a stationary observer as running at one-half its normal rate?

Answer: 0.866 c
ES   Moderate

26.39: A spaceship visits Alpha Centauri and returns to earth. Alpha Centauri is 4.5 light-years from earth (our second closest star). If the spaceship travels at one-half the speed of light for essentially all of its expedition, how long was the ship gone according to an observer on the earth?

Answer: 18 yrs
ES   Easy

26.40: A spaceship visits Alpha Centauri and returns to earth. Alpha Centauri is 4.5 light-years from earth (our second closest star). If the spaceship travels at one-half the speed of light for essentially all of its expedition, how long was the ship gone according to an observer on the spaceship?

Answer: 15.6 yr
ES   Moderate

26.41: Two spaceships approach each other along a straight line at a constant velocity of 0.988 c as measured by the captain of one of the ships. An observer on earth is able to measure the speed of only one of the ships as 0.900 c. From the point of view of the observer on earth, what is the speed of the other ship?

Answer: 0.794 c
ES   Moderate

26.42: Two spaceships are traveling through space at velocities of 0.6 c and 0.9 c, respectively, with respect to earth. If they are headed directly toward each other, what is their approach velocity, as measured by the captain of either ship?

Answer: 0.974 c
ES   Easy

26.43: A person in a rocket ship traveling past the earth at a speed of 0.5 c fires a laser gun in the forward direction. With what speed does an observer on earth see the light pulse travel?

Answer: 3×10^8 m/s
ES   Easy

26.44: Two spaceships are traveling through space at 0.6 c relative to the earth. If the ships are headed directly toward each other, what is their approach velocity, as measured by a person on either craft?
   a) 1.2 c   b) c
   c) 0.6 c   d) None of the above

Answer: (d)
MC   Easy

26.45: The theory of special relativity
a) is based on a complex mathematical analysis.
b) has not been verified by experiment.
c) does not agree with Newtonian mechanics.
d) does not agree with electromagnetic theory.

Answer: (c)
MC   Easy

26.46: Consider two spaceships, each traveling at 0.5 c in a straight line. Ship A is moving directly away from the sun and ship B is approaching the sun. The science officers on each ship measure the velocity of light coming from the sun. What do they measure for this velocity?
a) Ship A measures it as less than c, and ship B measures it as greater than c.
b) Ship B measures it as less than c, and ship A measures it as greater than c.
c) On both ships it is measured to be less than c.
d) On both ships it is measured to be exactly c.
e) On both ships it is measured to be greater than c.

Answer: (d)
MC   Easy

26.47: Relative to a stationary observer, a moving clock
a) always runs slower than normal.
b) always runs faster than normal.
c) keeps its normal time.
d) can do any of the above. It depends on the relative velocity between the observer and the clock.

Answer: (a)
MC   Moderate

26.48: As the velocity of your spaceship increases, you would observe
a) that your precision clock runs slower than normal.
b) that the length of your spaceship has decreased.
c) that your mass has increased.
d) all of the above.
e) none of the above.

Answer: (e)
MC   Moderate

26.49: You are riding in a spaceship that has no windows, radios, or other means for you to observe or measure what is outside. You wish to determine if the ship is stopped or moving at constant velocity. What should you do?
a) You can determine if the ship is moving by determining the apparent velocity of light.
b) You can determine if the ship is moving by checking your precision time piece. If it's running slow, the ship is moving.
c) You can determine if the ship is moving either by determining the apparent velocity of light or by checking your precision time piece. If its running slow, the ship is moving.
d) You should give up because you have taken on an impossible task.

Answer: (d)
MC   Easy

26.50: An object moves in a direction parallel to its length with a velocity that approaches the velocity of light. The length of this object, as measured by a stationary observer,
a) approaches infinity.   b) approaches zero.
c) increases slightly.   d) does not change.

Answer: (b)
MC   Moderate

26.51: An object moves in a direction parallel to its length with a velocity that approaches the velocity of light. The width of this object, as measured by a stationary observer,
a) approaches infinity.   b) approaches zero.
c) increases slightly.   d) does not change.

Answer: (d)
MC   Moderate

26.52: An object moves in a direction parallel to its length with a velocity that approaches the velocity of light. The mass of this object, as measured by a stationary observer,
a) approaches infinity.   b) approaches zero.
c) increases slightly.   d) does not change.

Answer: (a)
MC   Easy

26.53: A spear is thrown by you at a very high speed. As it passes, you measure its length at one-half its normal length. From this measurement, you conclude that the moving spear's mass must be
a) one-half its rest mass.   b) twice its rest mass.
c) four times its rest mass.   d) none of the above.

Answer: (b)
MC   Moderate

26.54: One of Einstein's postulates in formulating the special theory of relativity was that the laws of physics are the same in reference frames that
a) accelerate.
b) move at constant velocity with respect to an inertial frame.
c) oscillate.
d) are stationary, but not in moving frames.

Answer: (b)
MC   Easy

26.55: If the velocity of your spaceship goes from 0.3 c to 0.6 c, then your mass will increase by
a) 19%   b) 38%   c) 100%   d) 200%

Answer: (a)
MC   Moderate

26.56: How fast must something be traveling if its mass increases by 10%?

Answer: 1.25×10^8 m/s
ES   Moderate

26.57: If you were to measure your pulse rate while in a spaceship moving away from the sun at a speed close to the speed of light, you would find that it was
a) much faster than normal.
b) much slower than normal.
c) the same as it was here on earth.

Answer: (c)
MC   Moderate

26.58: The atomic bomb that was dropped on Nagasaki in 1945 killed 140,000 people, helping to end World War II on the next day. It released energy equivalent to that of 20,000 tons of TNT explosive. How much mass was converted to energy when this took place? (1000 tons ↔ 4.3×10^{12} J) Incidentally, modern H-bombs have energy yields 1000 times as much!

Answer: 0.96 g
ES   Easy

26.59: A person of initial mass 70 kg climbs a stairway, rising 6 m in elevation. By how much does his mass increase by virtue of his increased potential energy?

Answer: 4.57×10^{-14} kg
ES   Moderate

26.60: Which of the following depends on the observer's frame of reference?
  a) The mass of the proton
  b) The length of a meter stick
  c) The half-life of a muon
  d) All of the above

Answer: (d)
  MC   Easy

26.61: George Gamow, the creator of the Big-Bang theory of the origin of the universe, asks you to imagine what would happen if the speed of light was 15 mi/h. If you were on the sidewalk looking at a bicyclist ride down the street, how does she look to you while riding, compared to when she stops?
  a) taller, the same mass, but stretched out like a limousine.
  b) shorter, the same mass, but stretched out like a limousine.
  c) same height, greater mass, and flatter in the direction she is moving.
  d) same height, smaller mass, and flatter in the direction she is moving.
  e) shorter, greater mass, but stretched out like a limousine.

Answer: (c)
  MC   Moderate

26.62: The total energy of a particle at rest is zero.

Answer: False
  TF   Easy

26.63: Consider a particle of mass m and rest mass m_0. Which of the following is the correct expression for the kinetic energy of such a particle?
  a) $m_0v^2/2$ b) $mv^2/2$
  c) $mc^2 - m_0c^2$ d) $1/2(mc^2 - m_0c^2)$

Answer: (c)
  MC   Moderate

26.64: How many joules of energy are required to accelerate one kilogram of mass from rest to a velocity of 0.866 c?
  a) 1.8×10^{17} J   b) 9×10^{16} J
  c) 3×10^3 J   d) None of the above

Answer: (b)
  MC   Moderate

26.65: The amount of energy equivalent to one kilogram of mass at rest is
a) 9×10^{16} J
b) 3×10^8 J
c) 4.5×10^{16} J
d) None of the above.

Answer: (a)
MC Easy

26.66: What happens to the kinetic energy of a speedy proton when its relativistic mass doubles?
a) It doubles.
b) It more than doubles.
c) It less than doubles.
d) It must increase, but it is impossible to say by how much.

Answer: (b)
MC Moderate

26.67: What happens to the total relativistic energy of a speedy proton when its relativistic mass doubles?
a) It doubles.
b) It more than doubles.
c) It less than doubles.
d) It must increase, but it is impossible to say by how much.

Answer: (a)
MC Easy

26.68: An electron is accelerated through 100 kV. By what factor has its mass increased with respect to its rest mass?
a) 1.20
b) 1.55
c) 4.25
d) 8.00

Answer: (a)
MC Moderate

26.69: An electron has a relativistic momentum of 1.1×10^{-21} kg-m/s. What fraction of its total energy is its kinetic energy?

Answer: 0.759
ES Difficult

27. Early Quantum Theory and Models of the Atom

27.1: An electron is moving about a single proton in an orbit characterized by n = 4. How many of the electron's de Broglie wavelengths fit into the circumference of this orbit?
   a) 1   b) 2   c) 4   d) 16

Answer: (c)
MC   Easy

27.2: A beam of X-rays of frequency f is incident upon a substance that scatters the beam in various directions. If we measure the frequency of the scattered X-rays, we will find
   a) X-rays with frequency less than f.
   b) X-rays with frequency greater than f.
   c) only X-rays with frequency f.
   d) X-rays with frequencies ranging from less than f to greater than f.

Answer: (a)
MC   Moderate

27.3: A beam of X-rays of wavelength λ is incident upon a substance that scatters the beam in various directions. If we measure the wavelength of the scattered X-rays, we will find
   a) X-rays with wavelength less than λ.
   b) X-rays with wavelength greater than λ.
   c) only X-rays with wavelength λ.
   d) X-rays which range in wavelength from less than λ to greater than λ.

Answer: (b)
MC   Moderate

27.4: According to the Bohr model of the atom, the angular momentum of an electron around the nucleus
   a) could equal any positive value.
   b) must equal an integral multiple of h.
   c) must equal an integral multiple of $h/2\pi$.
   d) decreases with time, eventually becoming zero.

Answer: (c)
MC   Easy

27.5: If r_1 is the smallest orbital radius around a single proton, then r_6 is equal to
   a) $36r_1$.   b) $12r_1$.   c) $6r_1$.   d) $2.45r_1$.

Answer: (a)
MC   Moderate

27.6: What is the de Broglie wavelength of a hydrogen atom's electron when it is in the state n = 3?

Answer: 9.97×10^{-10} m
ES   Moderate

27.7: What is the frequency of revolution of a hydrogen atom's electron when it is in the n = 3 state?

Answer: 2.20×10^{15} Hz
ES   Difficult

27.8: What transition in a hydrogen atom results in the emission of light of wavelength 1007 nm?

Answer: n = 7 to n = 3
ES   Difficult

27.9: What wavelength of a photon is emitted when a singly ionized helium atom makes a transition from n = 2 to n = 1?

Answer: 30.5 nm
ES   Moderate

27.10: How much energy is carried by a photon with frequency 150 GHz?
    a) 1.5×10^{-20} J   b) 1.0×10^{-22} J
    c) 9.9×10^{-23} J   d) 1.4×10^{-25} J

Answer: (c)
MC   Easy

27.11: What frequency of electromagnetic radiation has energy 4.7 x 10^{-25} J?
    a) 710 kHz   b) 4.7 MHz   c) 710 MHz   d) 1.4 GHz

Answer: (c)
MC   Easy

27.12: What frequency of electromagnetic radiation has an energy of 58.1 μeV?
    a) 1.4 MHz   b) 711 MHz   c) 7.1 GHz   d) 14 GHz

Answer: (d)
MC   Moderate

27.13: How much energy is carried by a photon with a frequency of 100,000 GHz?
    a) 4.73×10^{-42} J   b) 4.37×10^{-24} J
    c) 6.63×10^{-24} J   d) 6.63×10^{-20} J

Answer: (d)
MC   Easy

27.14: An electron is traveling at 20 m/s. What is its de Broglie wavelength?

Answer: 36.4 μm
        ES   Easy

27.15: An electron undergoes a transition from the n = 4 state to the n = 2 state. What is the frequency of the photon emitted?

Answer: 6.15 x 10^{14} Hz
        ES   Difficult

27.16: Electrons being emitted when light hits a metallic surface is an example of photon absorption.

Answer: False
        TF   Easy

27.17: A quantum of electromagnetic radiation is called a proton.

Answer: False
        TF   Easy

27.18: In the Bohr model, the angular momentum of the electron is quantized.

Answer: True
        TF   Easy

27.19: An atom in the ground state can gain energy by emitting a photon.

Answer: False
        TF   Easy

27.20: A hologram is a three-dimensional image made by a laser.

Answer: True
        TF   Easy

27.21: Albert Einstein was the first person to explain the blackbody spectrum.

Answer: False
        TF   Easy

27.22: Irwin Schoedinger was the first to explain the photoelectric effect.

Answer: False
        TF   Easy

27.23: The hydrogen spectrum was first explained by Neils Bohr.

Answer: True
TF   Easy

27.24: As the temperature of a blackbody decreases, the peak wavelength emitted increases.

Answer: True
TF   Easy

27.25: In a photoelectric effect experiment, the wavelength of the incident light is decreased. The energy of the emitted electrons decreases.

Answer: False
TF   Easy

27.26: In a photoelectric effect experiment, the intensity of the incident light decreases. The energy of the emitted electrons decreases.

Answer: False
TF   Easy

27.27: An electron in a hydrogen atom falls from the $n = 5$ state to the $n = 2$ state. What is the energy of the emitted photon?

Answer: 2.86 eV
ES   Moderate

27.28: What wavelength of light must be absorbed by a ground state hydrogen atom to raise it to the $n = 3$ state?

Answer: 103 nm
ES   Moderate

27.29: The units of Planck's constant are
   a) $kg \cdot m/s$ b) $kg \cdot m^2/s$ c) $kg \cdot m/s^2$ d) $kg \cdot m^2/s^2$

Answer: (b)
MC   Moderate

27.30: The energy of electrons emitted in the photoelectric effect does not depend on the work function.

Answer: False
TF   Moderate

27.31: A metal surface has a work function of 2.7 eV. The surface is illuminated by 450 nm light. What is the maximum energy of the emitted electrons?

Answer: 0.062 ev
ES Moderate

27.32: A metal surface has a work function of 2.7 eV. The surface is illuminated by 450 nm light. What is the maximum speed of the emitted electrons? (The mass of an electron is 9.11×10^{-31} kg.)

Answer: 1.48×10^5 m/s
ES Difficult

27.33: 450 nm light shines on a metal surface. Electrons are emitted with a maximum energy of 1.20 eV. What is the work function for the surface?

Answer: 1.56 eV
ES Moderate

27.34: A photon has a momentum of 2.3×10^{-25} kg·m/s. What is the photon's frequency?

Answer: 1.04×10^{17} Hz
ES Easy

27.35: A photon has a wavelength of 450 nm. What is its momentum?

Answer: 1.47×10^{-27} kg·m/s
ES Easy

27.36: A neuton (mass 1.67×10^{-27} kg) has a speed of 4000 km/s. What is its wavelength?

Answer: 9.93×10^{-14} m
ES Moderate

27.37: A car of mass 1500 kg moves at 36 km/h. What is its wavelength?

Answer: 4.4×10^{-38} m
ES Easy

27.38: A proton (mass 1.67×10^{-27} kg) has wavelength 2×10^{-14} m. What is its kinetic energy?

Answer: 2.06 MeV
ES Difficult

27.39: In a pair of accelerating plates, such as found inside a CRT, the electrons are emitted
  a) from the cathode which is positive, toward the anode which is positive.
  b) from the cathode which is negative, toward the anode which is positive.
  c) from the anode which is positive, toward the cathode which is negative.
  d) from the anode which is negative, toward the cathode which is positive.

Answer: (b)
  MC   Easy

27.1: **MATCH THE UNIT TO THE PHYSICAL QUANTITY.**

27.40: Wien constant

Answer: m-°K
  MA   Easy

27.41: Planck's constant

Answer: J-s
  MA   Easy

27.42: work function

Answer: J
  MA   Easy

27.43: Compton wavelength

Answer: m
  MA   Easy

27.44: Rydberg constant

Answer: m^{-1}
  MA   Easy

27.45: A blackbody is an ideal system that
a) absorbs 100% of the light incident upon it, but cannot emit light of its own (i.e., a "black" body).
b) emits 100% of the light it generates, but cannot absorb radiation of its own.
c) either absorbs 100% of the light incident upon it, or emits 100% of the radiation it generates.
d) absorbs 50% of the light incident upon it, and emits 50% of the radiation it generates.

Answer: (c)
MC   Easy

27.46: What is a photon?
a) An electron in an excited state
b) A small packet of electromagnetic energy that has particle-like properties
c) One form of a nucleon, one of the particles that makes up the nucleus
d) An electron that has been made electrically neutral
e) Another name for a neutrino

Answer: (b)
MC   Easy

27.47: A photon is a particle that
a) has zero electric charge.
b) has zero electric field associated with it.
c) cannot travel in a vacuum.
d) has a velocity in a vacuum that varies with the photon frequency.
e) is usually observed only in connection with nuclear reactions.

Answer: (a)
MC   Easy

27.48: Which of the following is an accurate statement?
a) In vacuum, ultraviolet photons travel faster than infrared photons.
b) Photons can have positive or negative charge.
c) An ultraviolet photon has more energy than an infrared photon.
d) Photons do not have momentum (i.e., they cannot exert pressure on things).

Answer: (c)
MC   Easy

27.49: What is the energy, in eV, of a photon of wavelength 550 nm?

Answer: 2.26 eV
ES   Easy

27.50: What is the wavelength of a photon of energy 2.0 eV?

Answer: 622 nm
        ES   Easy

27.51: A small gas laser of the type used in classrooms may radiate
       light at a power level of 2 mW. If the wavelength of the emitted
       light is 642 nm, how many photons are emitted per second?

Answer: 6.5×10^{15}
        ES   Moderate

27.52: The human eye can just detect green light of wavelength 500 nm,
       which arrives at the retina at the rate of 2×10^{-18} W. How many
       photons arrive each second?

Answer: 5
        ES   Moderate

27.53: A beam of red light and a beam of violet light each deliver the
       same power on a surface. For which beam is the number of photons
       hitting the surface per second the greatest?
       a) The red beam
       b) The violet beam
       c) The number of photons per second is the same for both beams.
       d) This cannot be answered without knowing just what the light
       intensity is.
       e) This cannot be answered without knowing just what the
       wavelengths of the light used are.

Answer: (a)
        MC   Moderate

27.54: Which color of light has the lowest energy photons?
       a) Red   b) Yellow   c) Green   d) Blue

Answer: (a)
        MC   Easy

27.55: The energy of a photon depends on
       a) its amplitude.   b) its velocity.
       c) its frequency.   d) none of the above.

Answer: (c)
        MC   Easy

27.56: The ratio of energy to frequency for a given photon gives
       a) its amplitude.   b) its velocity.
       c) Planck's constant.   d) its work function.

Answer: (c)
        MC   Easy

27.57: How much energy, in joules, is carried by a photon of wavelength 660 nm?
   a) 1.46×10^{-48} J
   b) 3.01×10^{-19} J
   c) 6.63×10^{-34} J
   d) None of the above

Answer: (b)
   MC   Easy

27.58: When the accelerating voltage in an X-ray tube is doubled, the minimum wavelength of the X-rays
   a) is increased to twice the original value.
   b) is increased to four times the original value.
   c) is decreased to one-half the original value.
   d) is decreased to one-fourth the original value.

Answer: (c)
   MC   Moderate

27.59: What are the shortest wavelength X-rays produced by a 50,000-V X-ray tube?

Answer: 2.49×10^{-11} m
   ES   Easy

27.60: A metallic surface has a work function of 2.5 eV. What is the longest wavelength that will eject electrons from the surface of this metal?

Answer: 497 nm
   ES   Easy

27.61: Planck's constant
   a) sets an upper limit to the amount of energy that can be absorbed or emitted.
   b) sets a lower limit to the amount of energy that can be absorbed or emitted.
   c) relates mass to energy.
   d) does none of the above.

Answer: (b)
   MC   Moderate

27.62: A metallic surface is illuminated with light of wavelength 400 nm. If the work function for this metal is 2.4 eV, calculate the maximum kinetic energy of the ejected electrons, in electron-volts.

Answer: 0.71 eV
   ES   Easy

27.63: At what rate are photons emitted by a 50-W sodium vapor lamp? (Assume that the lamp's light is monochromatic and of wavelength 589 nm.)

Answer: 1.48×10^{20} per second
ES   Moderate

27.64:

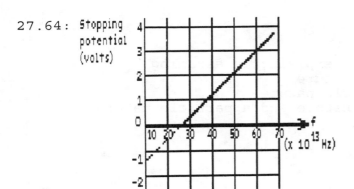

The graph shown is a plot based on student data from their testing of a photoelectric material. Determine (a) the cutoff frequency. (b) the work function.

Answer: (a) 25×10^{13} Hz   (b) 1.7×10^{-19} J
ES   Moderate

27.65: Molybdenum has a work function of 4.2 eV. What is the kinetic energy of the electrons emitted in the photoelectric effect when illuminated with light of wavelength 249 nm?

Answer: 0.8 eV
ES   Easy

27.66: A metal surface is illuminated with blue light and electrons are ejected at a given rate each with a certain amount of energy. If the intensity of the blue light is increased, electrons are ejected
a) at the same rate, but with more energy per electron.
b) at the same rate, but with less energy per electron.
c) at an increased rate with no change in energy per electron.
d) at a reduced rate with no change in energy per electron.

Answer: (c)
MC   Moderate

27.67: In the photoelectric effect, the energies of the ejected electrons
a) are proportional to the speed of light.
b) are proportional to the intensity of light.
c) are proportional to the frequency of light.
d) vary randomly.

Answer: (c)
MC   Moderate

27.68: In order for a photon to eject an electron from a metal's surface
in the photoelectric effect, the photon's
a) frequency must be greater than a certain minimum value.
b) speed must be greater than a certain minimum value.
c) wavelength must be greater than a certain minimum value.
d) momentum must be zero.

Answer: (a)
MC   Easy

27.69: The photoelectric effect is explainable assuming
a) that light has a wave nature.
b) that light has a particle nature.
c) that light has a wave nature _and_ a particle nature.
d) none of the above.

Answer: (b)
MC   Easy

27.70: X-rays with a wavelength of 0.00100 nm are scattered by free
electrons at 130°. What is the kinetic energy of each recoil
electron?

Answer: 994 keV
ES   Difficult

27.71: In a Compton scattering experiment, what scattering angle
produces the greatest change in wavelength?
a) Zero degrees   b) 90°
c) 180°   d) None of the above

Answer: (c)
MC   Moderate

27.72: In the Compton effect, as the scattering angle increases, the
frequency of the X-rays scattered at that angle
a) increases.   b) decreases.
c) does not change.   d) varies randomly.

Answer: (b)
MC   Moderate

27.73: What change in wavelength is expected when X-rays are scattered
at 90°?
a) No change at this angle   b) 4.84×10^{-3} nm
c) 2.42×10^{-3} nm   d) 1.18×10^{-3} nm

Answer: (c)
MC   Moderate

27.74: The energy required to ionize a hydrogen atom is
   a) 27.2 eV   b) 13.6 eV   c) 6.8 eV   d) 3.4 eV

Answer: (b)
   MC Easy

27.75: What is the radius of the first electron orbit in a singly ionized helium ion?

Answer: 0.026 nm
   ES Moderate

27.76: What is the energy, in eV, of the $n = 3$ state in hydrogen?

Answer: -1.51 eV
   ES Easy

27.77: What is the ionization energy for a doubly ionized lithium ion ($Z = 3$)?

Answer: 122 eV
   ES Moderate

27.78: When an electron jumps from an orbit where $n = 4$ to one where $n = 2$
   a) a photon is emitted.   b) a photon is absorbed.
   c) two photons are emitted.   d) two photons are absorbed.
   e) none of the above occur.

Answer: (a)
   MC Easy

27.79: The distance between adjacent orbit radii in a hydrogen atom
   a) increases with increasing values of n.
   b) decreases with increasing values of n.
   c) remains constant for all values of n.
   d) varies randomly with increasing values of n.

Answer: (a)
   MC Moderate

27.80: The energy difference between adjacent orbit radii in a hydrogen atom
   a) increases with increasing values of n.
   b) decreases with increasing values of n.
   c) remains constant for all values of n.
   d) varies randomly with increasing values of n.

Answer: (b)
   MC Moderate

27.81: In state n = 1, the energy of the hydrogen atom is -13.58 eV. What is its energy in state n = 2?
   a) -6.79 eV   b) -4.53 eV   c) -3.40 eV   d) -1.51 eV

Answer: (c)
   MC   Easy

27.82: In making a transition from state n = 1 to state n = 2, the hydrogen atom must
   a) absorb a photon of energy 10.2 eV.
   b) emit a photon of energy 10.2 eV.
   c) absorb a photon of energy 13.58 eV.
   d) emit a photon of energy 13.58 eV.

Answer: (a)
   MC   Moderate

27.83: What is the ionization energy of the neutral hydrogen atom?
   a) 27.2 eV   b) 13.6 eV
   c) 6.8 eV   d) None of the above

Answer: (b)
   MC   Easy

27.84: What is the ionization energy of singly ionized helium?
   a) 54.4 eV   b) 27.2 eV
   c) 13.6 eV   d) None of the above

Answer: (a)
   MC   Easy

27.85: In order for the atoms in a neon discharge tube (or a neon beer sign) to emit their characteristic orange light, it is necessary that
   a) each atom carry a net electric charge.
   b) the atoms be continually replaced with fresh atoms, because the energy of the atoms tends to be used up with continued excitation, resulting in dimmer and dimmer light.
   c) there be no unoccupied energy levels in each atom.
   d) the neon nucleus be unstable.
   e) electrons first be given energy to raise them from their ground state to an excited state.

Answer: (e)
   MC   Moderate

27.86: In a famous experiment done at the end of the 19th century, two
metal electrodes were placed in an evacuated glass tube. A high
voltage was applied between them. By using an appropriate metal
for the cathode, (e.g. sodium or potassium), it was found that a
current could be made to flow when the cathode was illuminated
with blue light, but no current would flow when red light was
used, no matter how bright the light. Einstein was able to
explain this strange phenomenon, and as a result, this experiment
was considered the first experimental demonstration of
a) the particle nature of light.
b) the wave nature of electrons.
c) the wave nature of light.
d) the particle nature of electrons.
e) the equivalency of electron and photons.

Answer: (a)
MC   Easy

27.87: You may have heard it said that all objects, even ones at the
temperature of our bodies, are continually emitting
electromagnetic radiation.   Is this true?
a) Yes, and the radiation referred to results from the trace
amounts of radioactive material that have accumulated in our
bodies from the food we eat.
b) Yes, and the radiation from our bodies is mostly in the
infrared region of the spectrum, and hence not detected by our
eyes.
c) Yes, and this radiation is mostly in the form of radio waves,
which is a common source of static often heard on an AM radio.
d) No, this is not true. If a person were emitting
electromagnetic radiation, he could be seen glowing in the dark,
and this is obviously not the case for most people.
e) No, this is not true, because our bodies are electrically
neutral.

Answer: (b)
MC   Easy

27.88: Bound electrons have negative energy.

Answer: True
TF   Easy

27.89: Photons are emitted when a bound electron jumps up to a higher
energy state.

Answer: False
TF   Easy

27.90: Atoms in crystals are typically separated by distances of 0.1 nm.
What KE must an electron have, in eV, in order to have a
wavelength of 0.1 nm?

Answer: 151 eV
ES   Moderate

27.91: As a particle travels faster, its de Broglie wavelength
      a) increases.
      b) decreases.
      c) remains constant.
      d) could be any of the above; it depends on other factors.

Answer: (b)
      MC   Moderate

27.92: When a photon is scattered from an electron, there will be an
      increase in its
      a) energy.   b) frequency.
      c) wavelength.   d) momentum.

Answer: (c)
      MC   Moderate

27.93: When an electron is accelerated to a higher speed, there is a
      decrease in its
      a) energy.   b) frequency.
      c) wavelength.   d) momentum.

Answer: (c)
      MC   Moderate

27.94: The reason the wavelike nature of a moving baseball is not
      noticed in everyday life is that
      a) it doesn't have a wavelike nature.
      b) its wavelength is too small.
      c) its frequency is too small.
      d) its energy is too small.
      e) no one pays attention to such things except for the Mets; and
      they can't hit a curve ball anyway.

Answer: (b)
      MC   Easy

27.95: The part of an electron microscope that plays the same role as
      the lenses do in an optical microscope is
      a) the vacuum chamber.   b) the coils.
      c) the cathode.   d) the deflector plates.

Answer: (b)
      MC   Easy

27.96: Which of the following microscopes is capable of "photographing"
      individual atoms?
      a) Light microscope
      b) Scanning tunneling microscope
      c) Transmission electron microscope
      d) Scanning electron microscope

Answer: (b)
      MC   Moderate

27.97: What is the de Broglie wavelength of a ball of mass 200 g moving
with a speed of 30 m/s?
a) 1.1×10^{-34} m
b) 2.2×10^{-34} m
c) 4.5×10^{-28} m
d) 6.67×10^{-27} m

Answer: (a)
MC   Easy

27.98: An electron has a wavelength of 0.123 nm. What is its energy in
eV? (This energy is not in the relativistic region.)
a) 20 eV   b) 60 eV   c) 80 eV   d) 100 eV

Answer: (d)
MC   Moderate

27.99: A person of mass 50 kg has a wavelength of 4.42×10^{-36} m when
running. How fast is she running?
a) 2 m/s   b) 3 m/s   c) 4 m/s   d) 5 m/s

Answer: (b)
MC   Moderate

27.100: Although de Broglie had not yet made his discovery of the
wavelength associated with electrons at the time Bohr advanced
his model of the atom, one can see how such an idea fits in with
Bohr's ideas concerning electron orbits. The basic idea is that
a) since electrons are waves, they can be anywhere in the atom.
b) the wave associated with an electron must interfere
constructively with itself, which suggests an orbit
circumference should be a multiple of the electron wavelength.
c) the diameter of the orbit will be equal to the electron
wavelength.
d) since electrons are waves, their frequency of revolution will
be given by v/λ, where v is the electron velocity in an orbit.
e) electrons would be expected to emit light of well-defined
sharp wavelengths, since they themselves have definite
wavelengths.

Answer: (b)
MC   Moderate

27.101: What advantage might an electron microscope have over a light
microscope?
a) Electrons are more powerful.
b) Shorter wavelengths are possible.
c) Longer wavelengths are possible.
d) None.

Answer: (b)
MC   Easy

27.102: Suppose that a particle is confined in a one-dimensional box that extends from the origin to x = a. If the wave function of the particle is given by Ax, where A is a constant, what is the probability of finding the particle in the right half of the box?

Answer: 75%
   ES   Difficult

27.103: In a pair of accelerating plates, such as found inside a CRT, the electrons are emitted
   a) from the cathode which is positive, toward the anode which is positive.
   b) from the cathode which is negative, toward the anode which is positive.
   c) from the anode which is positive, toward the cathode which is negative.
   d) from the anode which is negative, toward the cathode which is positive.

Answer: (b)
   MC   Easy

28. Quantum Mechanics of Atoms

28.1: An electron is known to be confined to a region of width 0.1 nm. What is an approximate expression for the least kinetic energy it could have, in eV?

Answer: 3.8 eV
ES   Moderate

28.2: In an actual hydrogen atom, an electron in the lowest state has an angular momentum
a) greater than h.
b) equal to h.
c) equal to $h/2\pi$.
d) equal to zero.

Answer: (d)
MC   Moderate

28.3: A neutral atom has an electron configuration of $1s^2 2s^2 2p^6 3s^2 3p^2$. What is its atomic number?
a) 5   b) 11   c) 14   d) 20

Answer: (c)
MC   Easy

28.4: In terms of an atom's electron configuration, the letters K, L, M, and N refer to
a) different shells with n equal to 1, 2, 3, or 4 respectively.
b) different subshells with l equal to 1, 2, 3, or 4 respectively.
c) the four possible levels for the magnetic quantum number.
d) the four possible quantum nummbers.

Answer: (a)
MC   Moderate

28.5: A neutral atom has an electron configuration of $1s^2 2s^2 2p^6$. If a neutral atom holds one additional electron, what is the ground state configuration?
a) $1s^2 2s^2 2p^6 3s^1$
b) $1s^2 2s^2 2p^7$
c) $1s^2 2s^3 2p^6$
d) None of the above.

Answer: (a)
MC   Easy

28.6: The Pauli exclusion principle states that
a) there are only four quantum numbers: n, l, m_l, and m_s.
b) no two quantum numbers can have the same value.
c) no two electrons can occupy the same quantum state.
d) no two quantum numbers can equal the same number.

Answer: (c)
MC   Moderate

28.7: In the ground state, the quantum numbers (n, l, m_l, m_s) for hydrogen are, respectively,
   a) 1, 1, 1, 1
   b) 1, 0, 0, 0
   c) 1, 0, 0, ±1/2
   d) 1, 1, 1, ±1/2

Answer: (c)
   MC   Moderate

28.8: Consider ground-state helium holding two electrons in orbit. If one of the electrons has quantum numbers (n, l, m_l, m_s) of 1, 0, 0, -1/2 respectively, the quantum numbers for the other electron will be
   a) 1, 1, 0, -1/2
   b) 1, 0, 0, +1/2
   c) 1, 1, 1, +1/2
   d) None of the above.

Answer: (b)
   MC   Moderate

28.9: The wave equation for hydrogen has solutions only if the three quantum numbers n, l, and m_l meet certain conditions. One of these conditions specifies that n
   a) can be any real number.
   b) can be any non-negative integer.
   c) can be any integer.
   d) can be any positive integer.

Answer: (d)
   MC   Moderate

28.10: The wave equation for hydrogen has solutions only if the three quantum numbers n, l, and m_l meet certain conditions. One of these conditions specifies than l
   a) is either zero or +1.
   b) is either equal to or less than n-1.
   c) is a positive integer.
   d) has an absolute value that is either equal to or less than n.

Answer: (b)
   MC   Moderate

28.11: The wave equation for hydrogen has solutions only if the three quantum numbers n, l, and m_l meet certain conditions. One of these conditions specifies that m_l
   a) has an absolute value either equal to or less than l.
   b) is equal to or less than n.
   c) is equal to or greater than 1.
   d) can be any integer.

Answer: (a)
   MC   Moderate

28.12: The reason the position of a particle cannot be specified with
infinite precision is the
a) exclusion principle.   b) uncertainty principle.
c) photoelectric effect.   d) principle of relativity.

Answer: (b)
MC   Easy

28.13: The concept of a wave function was developed by Neils Bohr.

Answer: False
TF   Moderate

28.14: The uncertainty principle was developed by Werner Heisenberg.

Answer: True
TF   Moderate

28.15: Irwin Schroedinger is generally acknowledged as the sole primary
founder of quantum mechanics.

Answer: False
TF   Moderate

28.16: An atomic X-ray spectrum has characteristic peaks because of the
photoelecrtric effect.

Answer: False
TF   Moderate

28.17: An atomic X-ray spectrum has a continuous background because the
incoming electrons decelerate rapidly.

Answer: True
TF   Moderate

28.18: Laser light is coherent.

Answer: True
TF   Easy

28.19: Lasers work better if there is a population inversion of
electrons, but it is not necessary.

Answer: False
TF   Moderate

28.20: Laser light is a single wavelength.

Answer: True
TF   Moderate

28.21: A baseball has mass 143 g and speed 45.0 m/s, with the speed known to within 0.1%. What is the minimum uncertainty in the position of the baseball?

Answer: 1.65×10^{-32} m
ES Moderate

28.22: An electron inside a hydrogen atom is confined to within a space of about 0.1 nm. If the electron mass is 9.11×10^{-31} kg, what is the uncertainty in the electron's velocity?

Answer: 1.16×10^6 m/s
ES Moderate

28.23: The Bohr model cannot account for the relative strength of various spectral lines.

Answer: True
TF Moderate

28.24: The Bohr model cannot predict the shape of electron orbitals.

Answer: True
TF Moderate

28.25: The Bohr model cannot predict the wavelengths of emitted light.

Answer: False
TF Moderate

28.26: In a hydrogen atom, a given electron has n = 7. How many values can l have?
a) 6   b) 7   c) 15   d) 98

Answer: (b)
MC Moderate

28.27: In a hydrogen atom, a given electron has l = 7. How many values can m_l have?
a) 6   b) 7   c) 15   d) 98

Answer: (c)
MC Moderate

28.28: In a hydrogen atom, an electron with n = 7 can exist in how many different quantum states?
a) 6   b) 7   c) 15   d) 98

Answer: (d)
MC Moderate

28.29: An atom has electron configuration $1s^2 2s^2 2p^6 3s^2 3p^2$. What element is this?
   a) Carbon   b) Nitrogen   c) Silicon   d) Germanium

Answer: (c)
   MC   Easy

28.30: How many electrons will fit into a 4f subshell?
   a) 3   b) 4   c) 7   d) 14

Answer: (d)
   MC   Easy

28.31: The values of n and l for a 4f subshell are
   a) n = 4, l = 4   b) n = 4, l = 3
   c) n = 3, l = 3   d) n = 4, l = 2

Answer: (b)
   MC   Easy

28.32: The word LASER is an acronym for
   a) Light Altered Spectra of Energy Radiated.
   b) LAtent Source of Enhanced Radiation.
   c) Light Amplification by the Stimulated Emission of Radiation.
   d) Light Absorbed States of Energetic Resonance.

Answer: (c)
   MC   Easy

28.33: In order to produce a hologram, one needs, in addition to an object and a piece of photographic film,
   a) a beam of monochromatic light and a mirror.
   b) a beam of monochromatic light and a lens.
   c) a beam of coherent light and a lens.
   d) a beam of coherent light and a mirror.

Answer: (d)
   MC   Easy

28.34: When a hologram is illuminated with a beam of coherent light, it produces
   a) both a real and a virtual image.
   b) only a real image of the object.
   c) only a virtual image of the object.
   d) none of the above.

Answer: (a)
   MC   Moderate

28.35: A hologram shows a magnifying glass held over a postage stamp. You accidentally drop the hologram, and it shatters into many small pieces. The magnifier and stamp will be 100% visible in any fragment you look at.

Answer: True
TF   Moderate

28.1: **MATCH THE DESCRIPTION TO THE PHYSICIST.**

28.36: Erwin Schrodinger

Answer: developed a wave equation for matter waves (1926)
MA   Moderate

28.37: Louis de Broglie

Answer: suggested the existence of matter waves (1924)
MA   Moderate

28.38: J. J. Thomson

Answer: discovered the electron (1897)
MA   Moderate

28.39: Albert Einstein

Answer: explained the photoelectric effect (1905)
MA   Moderate

28.40: Wolfgang Pauli

Answer: proposed the exclusion principle
MA   Moderate

28.41: C. J. Davisson & L. H. Germer

Answer: first to produce diffraction patterns of electrons in crystals (1927)
MA   Moderate

28.42: Werner Heisenberg

Answer: set the limits on the probability of measurement accuracy (1927)
MA   Moderate

28.43: The radius of a typical nucleus is about 5×10^{-15} m. Assuming this to be the uncertainty in the position of a proton in the nucleus, estimate the uncertainty in the proton's energy (in eV).

Answer: 40 MeV
ES Difficult

28.44: Suppose that the speed of an electron traveling 2000 m/s is known to an accuracy of 1 part in 10^5 (i.e., within 0.001%). What is the greatest possible accuracy within which we can determine the position of this electron?

Answer: 5.8 mm
ES Easy

28.45: Heisenberg's Uncertainty Principle states that
a) all measurements are to some extent inaccurate, no matter how good the instrument used.
b) we cannot in principle know simultaneously the position and momentum of a particle with absolute certainty.
c) we can never be sure whether a particle is a wave or a particle.
d) at times an electron appears to be a particle and at other times it appears to be a photon.
e) the charge on the electron can never be known with absolute accuracy.

Answer: (b)
MC Easy

28.46: Neutrons of energy 0.025 eV pass through two slits 1.0 mm apart. How far apart will peaks in beam intensity (due to constructive interference) be on a screen 0.5 m away?
a) 90 nm   b) 105 nm   c) 150 nm   d) 200 nm

Answer: (a)
MC Difficult

28.47: The uncertainty in the position of a proton is 0.053 nm. What is the uncertainty in its speed?
a) 1200 m/s   b) 975 m/s   c) 600 m/s   d) 2.2 m/s

Answer: (a)
MC Moderate

29. Molecules and Solids

29.1: A diatomic quantum mechanical oscillator has a moment of inertia of 7.73×10^{-45} kg·m^2. What is the rotational energy in the quantum state characterized by L = 2?

Answer: 2.72×10^{-5} eV
ES   Moderate

29.2: A diatomic molecule has 2.6×10^{-5} eV of rotational energy in the L = 2 quantum state. What is the rotational energy in the L = 1 quantum state?

Answer: 8.67×10^{-6} eV
ES   Easy

29.3: A diatomic molecule has 18×10^{-5} eV of rotational energy in the L = 2 quantum state. What is the rotational energy in the L = 0 quantum state?
a) 9×10^{-5} eV
b) 6×10^{-5} eV
c) 3×10^{-5} eV
d) 0 eV

Answer: (d)
MC   Easy

29.4: In its lowest quantum state, a diatomic quantum mechanical rotator has a rotational energy of
a) zero.
b) $h/2\pi I$
c) $h/\pi I$
d) None of the above.

Answer: (a)
MC   Easy

29.5: In general, which of the following is the strongest bond?
a) Hydrogen bond
b) Van der Waals bond
c) Ionic bond
d) Covalent bond

Answer: (c)
MC   Easy

29.6: In general, which of the following is the weakest bond?
a) Hydrogen bond
b) Van der Waals bond
c) Ionic bond
d) Covalent bond

Answer: (b)
MC   Easy

29.7: For a diatomic quantum mechanical rotator, the energy difference between adjacent energy levels
a) increases as L increases.
b) decreases as L increases.
c) is constant for all L.
d) varies randomly as L increases.

Answer: (a)
MC   Moderate

29.8: In its lowest quantum state, the energy of a diatomic harmonic oscillator is
a) (1/4) hf   b) (1/2) hf   c) hf   d) (3/2) hf

Answer: (b)
MC   Easy

29.9: A diatomic molecule is vibrating in the $\nu = 1$ quantum state with a frequency of 2×10^{13} Hz. What is the energy of vibration?
a) 0.041 eV   b) 0.083 eV   c) 0.124 eV   d) 0.166 eV

Answer: (c)
MC   Easy

29.10: A diatomic quantum mechanical rotator in the L = 1 quantum state has energy E. The same rotator in the L = 2 quantum state will have energy
a) 2 E   b) 3 E
c) 6 E   d) None of the above.

Answer: (b)
MC   Easy

29.11: For a diatomic quantum mechanical vibrator, the energy difference between adjacent quantum states
a) increases as ν increases.
b) decreases as ν increases.
c) is constant for all values of ν.
d) varies randomly as ν increases.

Answer: (c)
MC   Moderate

29.12: A diatomic quantum mechanical vibrator in its ground state has energy E. This same vibrator in its third state has energy
a) E   b) 3 E   c) 5 E   d) 7 E

Answer: (c)
MC   Moderate

29.13: When atoms combine to form stable molecules,
a) energy is absorbed by the molecules.
b) energy is released by the molecules.
c) no energy is transferred.
d) any of the above are possible, depending on the circumstances.

Answer: (b)
MC   Easy

29.14: Van der Waals bonding is caused by
a) the sharing of electrons between atoms.
b) the transfer of electrons between atoms.
c) unequal charge distribution around neutral molecules.
d) atoms bonding to hydrogen molecules.

Answer: (c)
MC   Easy

29.15: Ionic bonding is caused by
a) the sharing of electrons between atoms.
b) the transfer of electrons between atoms.
c) unequal charge distributions around neutral molecules.
d) atoms bonding to hydrogen molecules.

Answer: (b)
MC   Easy

29.16: Covalent bonding is caused by
a) the sharing of electrons between atoms.
b) the transfer of electrons between atoms.
c) unequal charge distributions around neutral molecules.
d) atoms bonding to hydrogen molecules.

Answer: (a)
MC   Easy

29.17: The energy gap between the valence and conduction bands in a certain semiconductor is 1.25 eV. What is the threshold wavelength for optical absorption in this substance?

Answer: 995 nm
ES   Moderate

29.18: Estimate the rotational energy (in eV) for a diatomic hydrogen molecule in the L = 2 quantum state. (The equilibrium separation for the H_2 molecule is 0.075 nm.)

Answer: 4.44×10^{-2} eV
ES   Difficult

29.19: An n-type semiconductor is produced by
   a) doping the host crystal with donor impurities.
   b) doping the host crystal with acceptor impurities.
   c) pure crystals of germanium.
   d) None of the above.

Answer: (a)
   MC   Easy

29.20: A p-type semiconductor is produced by
   a) doping the host crystal with donor impurities.
   b) doping the host crystal with acceptor impurities.
   c) pure crystals of germanium.
   d) None of the above.

Answer: (b)
   MC   Easy

29.21: When a voltage is applied across a p-type semiconductor, the holes
   a) are destroyed.
   b) move toward the positive electrode.
   c) move toward the negative electrode.
   d) do not move.

Answer: (c)
   MC   Easy

29.22: In a p-type semiconductor, a hole is
   a) a donor atom.
   b) an extra electron supplied by a donor atom.
   c) a missing atom in the crystalline structure.
   d) a region where an electron is missing.

Answer: (d)
   MC   Easy

30. Nuclear Physics and Radioactivity

30.1: Two nuclei with the same number of protons but different numbers of neutrons are called isotopes of a given element.

Answer: True
TF Easy

30.2: When a nucleus undergoes beta decay it changes into another isotope of the same element.

Answer: False
TF Moderate

30.3: Isotope is the term used to describe a nucleus which is unstable and hence radioactive.

Answer: False
TF Easy

30.4: When nucleons join to form a stable nucleus, energy is
a) destroyed.   b) absorbed.
c) released.   d) not transferred.

Answer: (c)
MC Easy

30.5: When a β^- particle is emitted from an unstable nucleus, the atomic mass number of the nucleus
a) increases by 1.   b) decreases by 1.
c) does not change.   d) does none of the above.

Answer: (c)
MC Easy

30.6: The atomic mass unit is defined as
a) the mass of a proton.
b) the mass of an electron.
c) the mass of a hydrogen-1 atom.
d) one twelfth the mass of a carbon-12 atom.

Answer: (d)
MC Easy

30.7: Which of the atomic particles has the least mass?
a) electron   b) proton   c) neutron   d) nucleon

Answer: (a)
MC Easy

30.8: An element with atomic mass number of 14 and atomic number 6 has how many neutrons?
    a) 6   b) 8   c) 14   d) 20

Answer: (b)
MC   Easy

30.9: An element with atomic mass number of 14 and atomic number 6 has how many protons?
    a) 6   b) 8   c) 14   d) 20

Answer: (a)
MC   Easy

30.10: An alpha particle will be attracted to a
    a) gamma ray.   b) proton.
    c) positive charge.   d) negative charge.

Answer: (d)
MC   Easy

30.11: Alpha particles have an atomic mass equal to
    a) 1   b) 2   c) 4   d) 6

Answer: (c)
MC   Easy

30.12: An element with atomic number 88 goes through alpha decay. Its atomic number is now
    a) 80   b) 84   c) 86   d) 88

Answer: (c)
MC   Easy

30.13: An element with atomic number 6 undergoes β^- decay. It atomic number is now
    a) 7   b) 6   c) 5   d) 2

Answer: (a)
MC   Easy

30.14: If the half-life of a material is 45 years, how much will be left after 100 years?
    a) More than 1/2   b) Less than 1/2
    c) More than 1/4   d) Less than 1/4

Answer: (d)
MC   Easy

30.15: If 4×10^{18} atoms decay with a half-life of 2.3 years, how many are remaining after 3.7 years?
   a) 2.5×10^{18} b) 1.7×10^{18} c) 1.3×10^{18} d) 1.1×10^{18}

Answer: (c)
   MC   Moderate

30.16: How long would it take 4×10^{20} atoms to decay to 1×10^{19} atoms if their half-life was 14.7 years?
   a) 29.4 years   b) 58.8 years   c) 78.2 years   d) 147 years

Answer: (c)
   MC   Difficult

30.17: What is the probability that an atom will decay this year if its half-life is 247 years?
   a) 0.0017   b) 0.0028   c) 0.0171   d) 0.0711

Answer: (b)
   MC   Moderate

30.18: A Thallium source was certified at 10 μCi ten years ago. What is its activity now if the half-life is 3.7 years?
   a) 4.7 μCi   b) 3.3 μCi   c) 1.5 μCi   d) 1.0 μCi

Answer: (c)
   MC   Moderate

30.19: How long will it take a 4.5 Ci sample of material to reach an activity level of 0.14 Ci if the half-life is 435 years?
   a) 14,478 years   b) 3245 years
   c) 2178 years   d) 1993 years

Answer: (c)
   MC   Difficult

30.20: Uranium-239 undergoes alpha decay. What is the daughter nucleus?

Answer: $^{235}_{90}$Th
   ES   Easy

30.21: Oxygen-19 undergoes β^- decay. What is the daughter nucleus?

Answer: $^{19}_{9}$F
   ES   Easy

30.22: Sodium-21 undergoes β^+ decay.   What is the daughter nucleus?

Answer: $^{21}_{10}$Ne

ES   Easy

30.23: Technetium-99m undergoes gamma decay.   What is the daughter nucleus?

Answer: $^{99}_{43}$Tc

ES   Easy

30.24: A radioactive isotope intially decays at a rate of 800 decays per second.   48 minutes later the rate is 100 counts per second. What is the half-life of the isotope?

Answer: 16 minutes
ES   Easy

30.25: What letter is used to designate atomic number?

Answer: Z
ES   Easy

30.26: What letter is used to designate atomic mass?

Answer: A
ES   Easy

30.27: An α particle is also known as
   a) an electron.   b) a positron.
   c) a helium nucleus.   d) a photon.

Answer: (c)
MC   Easy

30.28: A β^- particle is also known as
   a) an electron.   b) a positron.
   c) a helium nucleus.   d) a photon.

Answer: (a)
MC   Easy

30.29: A β^+ particle is also known as
   a) an electron.   b) a positron.
   c) a helium nucleus.   d) a photon.

Answer: (b)
MC   Easy

30.30: A gamma ray is also known as
    a) an electron.
    b) a positron.
    c) a helium nucleus.
    d) a photon.

Answer: (d)
    MC   Easy

30.31: Binding energy is the measure of how tightly the nucleons in a nucleus are bound together.

Answer: True
    TF   Easy

30.32: What is the binding energy per nucleon for Uranium-325?
    Given $m(^{235}_{92}U)$ = 235.043924 u

        $m(^{1}_{1}H)$ = 1.007825 u

        $m(^{1}_{0}n)$ = 1.008665 u

        1 u = 931.5 MeV

Answer: 7.59 MeV/nucleon
    ES   Moderate

30.33: The half-life of uranium-238 is 4.5 billion years, and the half-life of uranium-235 is 0.9 billion years. When the earth was formed 4.5 billion years ago, the ratio of U-238 to U-235 on Earth was 9:1. What is that ratio now?

Answer: 144:1
    ES   Difficult

30.34: In a $^{93}_{41}Nb$ nucleus, the number of protons, neutrons, and electrons is
    a) 41, 52, 93   b) 41, 52, 52   c) 41, 52, 41   d) 41, 52, 0

Answer: (d)
    MC   Easy

 30.1: **MATCH THE DESCRIPTION TO THE NAME OF THE PHYSICIST.**

30.35: Wolfgang Pauli

Answer: proposed the existence of the neutrino
    MA   Moderate   Table: 1

30.36: Marie & Pierre Curie

Answer: discovered radium and polonium
    MA   Moderate   Table: 1

30.37: Hans Geiger

Answer: invented particle detector tube
        MA   Moderate   Table: 1

30.38: Enrico Fermi

Answer: gave the neutrino its name
        MA   Moderate   Table: 1

30.39: Donald Glazer

Answer: invented the bubble chamber
        MA   Moderate   Table: 1

30.40: Ernest Rutherford

Answer: proposed the tiny central nucleus idea
        MA   Moderate   Table: 1

30.41: Henri Becquerel

Answer: discovered radioactivity
        MA   Moderate   Table: 1

30.42: Write the nuclear notation for the following nuclei: Hydrogen-2,
       Sulfur-33, and Lead-207.

Answer: $^2_1$H, $^{33}_{16}$S, $^{207}_{82}$Pb

        ES   Easy

30.43: Which of the following statements concerning the nuclear force is
       false?
       a) The nuclear force is very short-ranged.
       b) The nuclear force is very weak and much smaller in relative
       magnitude than the electrostatic and gravitational forces.
       c) The nuclear force is attractive and not repulsive.
       d) The nuclear force acts on both protons and neutrons.

Answer: (b)
        MC   Easy

30.44: Compared to the electrostatic force, the nuclear force between
       adjacent protons in a nucleus is
       a) much weaker.   b) about the same size.
       c) only slightly larger.   d) much larger.

Answer: (d)
        MC   Moderate

30.45: If an element of atomic number 15 has an isotope of mass number 32,
a) the number of neutrons in the nucleus is 15.
b) the number of neutrons in the nucleus is 17.
c) the number of protons in the nucleus is 17.
d) the number of nucleons in the nucleus is 15.

Answer: (b)
MC   Easy

30.46: Isotopes of an element have nuclei with
a) the same number of protons, but different numbers of neutrons.
b) the same number of protons, and the same number of neutrons.
c) a different number of protons, and a different number of neutrons.
d) a different number of protons, and the same number of neutrons.

Answer: (a)
MC   Easy

30.47: When a nucleus undergoes beta decay, it changes into another isotope of itself.

Answer: False
TF   Easy

30.48: All elements have the same number of isotopes.

Answer: False
TF   Easy

30.49: Not all isotopes are radioactive.

Answer: True
TF   Easy

30.50: The number of protons in an atom is
a) zero.
b) called the mass number.
c) equal to the number of neutrons.
d) equal to the number of electrons.
e) the same for all elements.

Answer: (d)
MC   Easy

30.51: The mass of an atom is
a) approximately equally divided between neutrons, protons, and electrons.
b) evenly divided between the nucleus and the surrounding electron cloud.
c) concentrated in the cloud of electrons surrounding the nucleus.
d) concentrated in the nucleus.
e) about 10^{-6} g.

Answer: (d)
MC   Easy

30.52: How many protons are there in the carbon-14 nucleus?
a) None   b) 1   c) 6   d) 8   e) 14

Answer: (c)
MC   Easy

30.53: Heavy-water molecules have a mass number of
a) 3   b) 10   c) 12   d) 18   e) 20

Answer: (e)
MC   Moderate

30.54: What is the approximate nuclear radius of an isotope of sodium with 11 protons and 12 neutrons?

Answer: 3.4×10^{-15} m
ES   Easy

30.55: From a knowledge of the size and mass of a nucleus, one can estimate the density of nuclear material. Such a calculation shows that the density of nuclear matter is on the order of magnitude of
a) 10^{17} kg/m^3 b) 10^{11} kg/m^3
c) 10^{8} kg/m^3 d) 10^{4} kg/m^3
e) 10^{2} kg/m^3

Answer: (a)
MC   Moderate

30.56: Atoms with the same atomic number but with different numbers of neutrons are referred to as
a) nucleons.   b) nuclides.
c) isotopes.   d) none of the above.

Answer: (c)
MC   Easy

30.57: An atom's mass number is determined by the number of
a) neutrons in its nucleus.
b) nucleons in its nucleus.
c) protons in its nucleus.
d) alpha particles in its nucleus.

Answer: (b)
MC   Easy

30.58: If an atom's atomic number is given by Z, its atomic mass by A,
and its neutron number by N, which of the following is correct?
a) N = A + Z
b) N = Z - A
c) N = A - Z
d) None of the above is correct.

Answer: (c)
MC   Easy

30.59: There is a limit to the size of a stable nucleus because of
a) the limited range of the strong nuclear force.
b) the weakness of the electrostatic force.
c) the weakness of the gravitational force.
d) none of the above.

Answer: (a)
MC   Moderate

30.60: When a gamma ray is emitted from an unstable nucleus,
a) the number of neutrons and the number of protons drop by two.
b) the number of neutrons drops by one and the number of protons
increases by one.
c) there is no change in either the number of neutrons or the
number of protons.
d) none of the above is correct.

Answer: (c)
MC   Easy

30.61: When an alpha particle is emitted from an unstable nucleus, the
atomic number of the nucleus
a) increases by 2.   b) decreases by 2.
c) increases by 4.   d) decreases by 4.
e) none of the above.

Answer: (b)
MC   Easy

30.62: When an alpha particle is emitted from an unstable nucleus, the atomic mass number of the nucleus
   a) increases by 2.   b) decreases by 2.
   c) increases by 4.   d) decreases by 4.
   e) none of the above.

Answer: (d)
   MC   Easy

30.63: When a β- particle is emitted from an unstable nucleus, the atomic number of the nucleus
   a) increases by 1.   b) decreases by 1.
   c) does not change.   d) none of the above.

Answer: (a)
   MC   Moderate

30.64: When a β+ particle is emitted from an unstable nucleus, the atomic number of the nucleus
   a) increases by 1.   b) decreases by 1.
   c) does not change.   d) none of the above.

Answer: (b)
   MC   Moderate

30.65: When a neutron is emitted from an unstable nucleus, the atomic mass number of the nucleus
   a) increases by 1.   b) decreases by 1.
   c) does not change.   d) none of the above.

Answer: (b)
   MC   Easy

30.66: During β- decay
   a) a neutron is transformed to a proton.
   b) a proton is transformed to a neutron.
   c) a neutron is ejected from the nucleus.
   d) a proton is ejected from the nucleus.

Answer: (a)
   MC   Easy

30.67: During β+ decay
   a) a neutron is transformed to a proton.
   b) a proton is transformed to a neutron.
   c) a neutron is ejected from the nucleus.
   d) a proton is ejected from the nucleus.

Answer: (b)
   MC   Easy

30.68: During internal conversion,
    a) a proton is absorbed by the nucleus.
    b) a neutron is absorbed by the nucleus.
    c) an electron is absorbed by the nucleus.
    d) an alpha particle is absorbed by the nucleus.

Answer: (c)
    MC Moderate

30.69: No charged particle is emitted as a result of internal conversion.

Answer: True
    TF Moderate

30.70: Complete the following nuclear decay equation: $^{210}_{84}Po* \rightarrow ^{210}_{84}Po$ + _____

Answer: γ radiation
    ES Moderate

30.71: Complete the following nuclear decay equation: $^{238}_{92}U \rightarrow ^{234}_{90}Th$ + _____

Answer: $^{4}_{2}He$ (α particle)
    ES Moderate

30.72: Complete the following nuclear decay equation: $^{22}_{11}Na$ + _____ $\rightarrow ^{22}_{10}Ne$

Answer: $^{0}_{-1}e$ (electron)
    ES Easy

30.73: The radioactivity due to carbon-14 measured in a piece of a wooden casket from an ancient burial site was found to produce 20 counts per minute from a given sample, whereas the same amount of carbon from a piece of living wood produced 160 counts per minute. The half-life of carbon-14, a beta emitter, is 5730 years. Thus we would estimate the age of the artifact to be about
    a) 5,700 years   b) 11,500 years
    c) 14,800 years   d) 17,200 years
    e) 23,000 years

Answer: (d)
    MC Moderate

30.74: A radioactive sample is determined to have 3.45×10^{19} nuclei present. Two days later (48 hours) it is again tested, but now found to have 2.00×10^{18} nuclei present. How many half-lives have elapsed during these 48 hours?

Answer: 4.11
ES Moderate

30.75: Suppose that in a certain collection of nuclei there were initially 1024 nuclei and 20 minutes later there was only one nuclei left, the others having decayed. On the basis of this information, how many nuclei would you estimate decayed in the first 6 minutes?

Answer: 896
ES Difficult

30.76: A radioactive substance with a half-life of 3 days has an initial activity of 0.24 Ci. What is its activity after 6 days?
　　a) 0.12 Ci　　　　　　　　　　b) 0.48 Ci
　　c) 0.06 Ci　　　　　　　　　　d) none of the above

Answer: (c)
MC Moderate

30.77: A radioactive substance containing 4.00×10^{16} unstable atoms has a half-life of 2 days. What is its activity (in curies) after 1 day?

Answer: 3.07 Ci
ES Difficult

30.78: In radioactive dating, carbon-14 is often used. This nucleus emits a single beta particle when it decays. When this happens, the resulting nucleus is
　　a) still carbon-14.　　　　　　b) boron-14.
　　c) nitrogen-14.　　　　　　　 d) carbon-15.
　　e) carbon-13.

Answer: (c)
MC Moderate

30.79: What happens to the half-life of a radioactive substance as it decays?
　　a) It remains constant.　　　　b) It increases.
　　c) It decreases.　　　　　　　d) It could do any of these.

Answer: (a)
MC Easy

30.80:

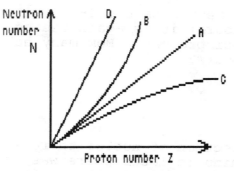

The neutron-proton stability curve of the isotopes of all elements will be plotted on a graph whose axes are shown.

The plot will look like
a) the straight line (N = Z), shown.
b) a curve above the line for N > Z.
c) a curve below the line for N < Z.
d) a straight line where N > Z.

Answer: (b)
MC Moderate

30.81: Stable nuclei with mass numbers greater than 40 have more neutrons than protons.

Answer: True
TF Moderate

30.82: Carbon-12 is unstable.

Answer: False
TF Moderate

30.83: Calculate the binding energy of ^{9}Be.

Answer: 58.2 MeV
ES Moderate

30.84: How much energy is required to remove one proton from ^{9}Be?

Answer: 16.9 MeV
ES Moderate

30.85: What is the binding energy of the ^{4}He nucleus?
a) 7.80 MeV   b) 14.15 MeV   c) 20.36 MeV   d) 28.3 MeV

Answer: (d)
MC Moderate

30.86: The binding energy per nucleon
a) increases steadily as we go to heavier elements.
b) decreases steadily as we go to heavier elements.
c) is approximately constant throughout the periodic table, except for very light nuclei.
d) has a maximum near iron in the periodic table.

Answer: (d)
MC   Moderate

30.87: The binding energy per nucleon is
a) directly proportional to atomic number.
b) inversely proportional to atomic number.
c) the same for all atoms.
d) none of the above.

Answer: (d)
MC   Moderate

30.88: The type of detector that uses a magnetic field to curve charged particles is a
a) Geiger tube.   b) scintillation counter.
c) cloud chamber.   d) bubble chamber.
e) spark chamber.

Answer: (d)
MC   Moderate

30.89: The type of detector that uses liquid hydrogen is a
a) Geiger tube.   b) scintillation counter.
c) cloud chamber.   d) bubble chamber.
e) spark chamber.

Answer: (d)
MC   Moderate

30.90: Cloud chambers have been replaced by bubble chambers because
a) the radioactive clouds were too dangerous to work with.
b) the density of fluids is greater than the density of vapors.
c) bubble chambers tend to be larger and more expensive.
d) gamma rays are visible in bubble chambers, but not in cloud chambers.

Answer: (b)
MC   Moderate

30.91: A carbon-14 nucleus decays to a nitrogen-14 nucleus by beta decay. How much energy (in MeV) is released if carbon-14 has a mass of 14.003242 u and nitrogen-14 has a mass of 14.003074 u?

Answer: 0.157 MeV
ES   Moderate

30.92: The expression $(M_X - M_Y - M_a) \times 931.5$ represents
a) the binding energy of the nucleus X.
b) the binding energy of the nucleus Y.
c) the energy released when nucleus X undergoes alpha decay.
d) the energy released when nucleus Y undergoes alpha decay.
e) the energy released in a typical fission reaction of uranium.

Answer: (c)
MC   Moderate

30.93: Find the energy in MeV released in the reaction

$$^{238}_{92}U \rightarrow {}^{234}_{90}Th + {}^{4}_{2}He$$

Mass of U-238 = 238.050786 u
Mass of Th-234 = 234.043583 u
Mass of He-4 = 4.002603 u

Answer: 4.28 MeV
ES   Moderate

30.94: Which of the following is most nearly the same as a gamma ray?
a) An alpha particle   b) A beta ray
c) Visible light   d) A proton
e) A neutron

Answer: (c)
MC   Easy

30.95: The existence of the neutrino was postulated in order to explain
a) alpha decay.   b) gamma emission.
c) beta decay.   d) fission.

Answer: (c)
MC   Moderate

30.96: Scientists were led to postulate the existence of the neutrino in order to
a) maintain the conservation of energy and of momentum in beta decay.
b) explain the intense radiation emitted by quasars.
c) account for alpha decay.
d) provide a mechanism for quark production.

Answer: (a)
MC   Moderate

30.97: In beta decay
   a) a proton is emitted.
   b) a neutron is emitted.
   c) an electron is emitted.
   d) an electron decays into another particle.

Answer: (c)
   MC   Easy

31. Nuclear Energy: Effects and Uses of Radiation

31.1: What is the mass of the products of a nuclear fission reaction compared to the mass of the original products?
a) greater
b) less
c) the same
d) varies according to the reaction

Answer: (b)
MC   Easy

31.2: What is the mass of the products of a nuclear fusion reaction compared to the original elements?
a) greater
b) less
c) the same
d) varies according to the reaction

Answer: (b)
MC   Easy

31.3: What is necessary to stop beta particles?
a) air alone   b) paper
c) metal foil   d) thick metal

Answer: (c)
MC   Easy

31.4: When lead-207 (Z = 82) is bombarded with neutrons, it can change into
a) lead-208   b) lead-206
c) tellurium-208 (Z = 81)   d) bismuth-208 (Z = 83)

Answer: (a)
MC   Easy

31.5: Determinining the composition of a substance by bombarding it with neutrons and measuring the radiation emitted is called nuclear emission analysis.

Answer: False
TF   Easy

31.6: Complete the equation:
$^{16}_{8}O + ^{4}_{2}He \rightarrow ^{19}_{10}Ne + ?$

Answer: $^{1}_{0}n$

ES   Moderate

31.7: Complete the equation:
$$^{38}_{19}K + ^{1}_{1}H \rightarrow ^{38}_{20}Ca + ?$$

Answer: $^{1}_{0}n$

ES Moderate

31.8: Complete the equation:
$$^{235}_{92}U \rightarrow ^{101}_{39}Y + ? + 3^{1}_{0}n$$

Answer: $^{131}_{53}I$

ES Moderate

31.9: Complete the equation:
$$^{3}_{1}H + ^{3}_{1}H \rightarrow ^{4}_{2}He + ?$$

Answer: $2^{1}_{0}n$

ES Moderate

31.10: In a nuclear explosion, each fission event produces, on average, 2.5 neutrons which can then produce further fissions. Suppose the time interval between the release of a neutron and the time it produces a fission event is 1×10^{-8} s. How long will it take to completely fission 15 kg of U-235?

Answer: $0.64 \mu s$
ES Difficult

31.11: In the sun, the net reaction which occurs is
$$4^{1}_{1}H \rightarrow ^{4}_{2}He + 2^{0}_{-1}e- + neutrinos$$
Given: $m(^{1}_{1}H) = 1.007825$ u

$m(^{4}_{2}He) = 4.002602$ u

$m(^{0}_{-1}e) = 0.000549$ u

1 u = 931.5 MeV
How much energy is released in this reaction?

Answer: 25.7 MeV
ES Moderate

31.12: It requires about 25 eV of energy to split an H_2O molecule into an H^+ radical and an OH^- radical. A gamma ray of wavelength 5 x 10^{-13} m is absorbed by tissue (which is mostly water). Approximately how many OH^- radicals are produced?
    a) 1,000   b) 10,000   c) 100,000   d) 1,000,000

Answer: (c)
       MC Moderate

31.13: What is created in pair production?
    a) A proton and an electron   b) A proton and an antiproton
    c) An electron and a positron   d) An electron and a photon

Answer: (c)
       MC Easy

31.1: **MATCH THE DESCRIPTION TO THE UNIT.**

31.14: becquerel (Bq)

Answer: SI unit of radioactivity (= 1 decay/s)
       MA Moderate

31.15: rad

Answer: 10^{-2} J/kg
       MA Moderate

31.16: roentgen (R)

Answer: 0.878 x 10^{-2} J/kg of dry air
       MA Moderate

31.17: curie (Ci)

Answer: 3.70 x 10^{10} decays/s
       MA Moderate

31.18: gray (Gy)

Answer: SI unit of absorbed dose = 1 J/kg = 100 rad
       MA Moderate

31.19: rem

Answer: effective dose (= (rads)(RBE))
       MA Moderate

31.20: sievert (Sv)

Answer: effective dose of 1 Gy
MA   Moderate

31.21: RBE

Answer: produces same tissue damage as 1 rad of gamma or X radiation
MA   Moderate

31.22: The chief hazard of radiation is
a) damage to living cells due to ionization.
b) damage to cells due to heating.
c) damage to living cells due to the creation of chemical impurities.
d) the creation of new isotopes within the body.

Answer: (a)
MC   Easy

31.23: A unit that measures the effective dose of radiation in a human is the
a) curie.   b) RBE.   c) rad.   d) rem.

Answer: (d)
MC   Moderate

31.24: All of the following are units used to describe radiation dosage in humans except
a) curie.   b) rad.
c) rem.   d) RBE.
e) sievert.

Answer: (a)
MC   Moderate

31.25: Complete the following nuclear reaction:

$^{16}_{8}O + ^{1}_{0}n \rightarrow$ _____ $+ ^{4}_{2}He$

Answer: $^{13}_{6}C*$
ES   Moderate

31.26: Complete the following nuclear reaction:

$^{27}_{13}Al \ (\alpha, n)$ _____

Answer: $^{30}_{15}P*$
ES   Moderate

31.27: A negative Q value for a reaction indicates the reaction is exoergic.

Answer: False
TF   Easy

31.28: Find the Q value of the following reaction:

$$_{7}^{14}N \quad + \quad _{2}^{4}He \quad \rightarrow \quad _{8}^{17}O \quad + \quad _{1}^{1}H$$
(14.003074 u)   (4.002603 u)   (16,999131 u)   (1.007825 u)

Answer: -0.001279 u
ES   Easy

31.29: The mass of a proton is 1.6726×10^{-27} kg and the mass of a neutron is 1.6749×10^{-27} kg. A proton captures a neutron forming a deuterium nucleus. One would expect the mass of this nucleus to be
a) equal to $(1.6726 + 1.6749) \times 10^{-27}$ kg.
b) less than $(1.6726 + 1.6749) \times 10^{-27}$ kg.
c) greater than $(1.6726 + 1.6749) \times 10^{-27}$ kg.
d) any of these; it depends on the energy released during the capture.

Answer: (b)
MC   Moderate

31.30: The Q value for a particular reaction is -2.4 MeV, and the reaction's threshold energy is 9.60 MeV. What is the ratio of the mass of the incident particle to the mass of the stationary target nucleus?
a) -0.75   b) 0.25   c) 3   d) 4   e) 5

Answer: (c)
MC   Moderate

31.31: When a target nucleus is bombarded by an appropriate beam of particles, it is possible to produce
a) a less massive nucleus, but not a more massive one.
b) a more massive nucleus, but not a less massive one.
c) a nucleus with smaller atomic number, but not one with a greater atomic number.
d) a nucleus with greater atomic number, but not one with a smaller atomic number.
e) a nucleus with either greater or smaller atomic number.

Answer: (e)
MC   Moderate

31.32: What is the energy of reaction of the process $^9_4\text{Be}(\alpha,n)^{12}_6\text{C}$?

(This reaction, first observed by Chadwick in 1930, led to his discovery of the neutron.)
a) 3.66 Mev
b) 5.60 MeV
c) 5.70 MeV
d) 6.11 MeV
e) 6.34 MeV

Answer: (c)
MC   Moderate

31.33: What is the source of the energy the sun radiates to us?
a) Chemical reactions
b) Nuclear fission reactions
c) Nuclear fusion reactions
d) Magnetic explosions
e) Cosmic rays

Answer: (c)
MC   Easy

31.34: A proton strikes an oxygen-18 nucleus producing fluorine-18 and another particle. What other particle is produced by this nuclear reaction?
a) A neutron
b) An alpha particle
c) A β- particle
d) A β+ particle

Answer: (a)
MC   Easy

31.35: Which of the following best describes the process in which energy is released in a conventional nuclear reactor?
a) The radiation given off by a naturally radioactive substance, uranium, is collected and used to make steam.
b) Uranium is reacted with oxygen in a combustion process that releases large amounts of radioactivity and heat.
c) Deuterium and tritium are joined together to form helium.
d) Uranium, when bombarded by neutrons, splits into fragments and releases two or three neutrons, and these neutrons in turn strike more uranium nuclei that split, thereby setting off a chain reaction that releases energy.
e) A uranium nucleus is energized to an excited state by neutron irradiation, and it then decays by emitting beta rays and gamma rays that heat water and create steam.

Answer: (d)
MC   Easy

31.36: Which of the following (if any) statements is true concerning a nuclear reactor as compared to a coal-fired power plant for generating electricity?
a) Higher voltages are generated when nuclear reactions are used.
b) The electricity from a nuclear reactor is slightly radioactive, whereas that from a coal-fired plant is not.
c) Nuclear energy can be converted directly into electricity, whereas a coal-fired plant must first generate steam, which in turn drives a turbine.
d) The radioactivity in the smoke from a coal-fired plant is comparable to that from a nuclear plant, but since it is so widely dispersed it does not present a major environmental hazard.
e) None of these statements is true.

Answer: (e)
MC   Easy

31.37: When light elements such as hydrogen undergo fusion,
a) there is a loss of mass.
b) there is an·increase in mass.
c) there is no change in mass.
d) electric charge can be annihilated.
e) more than one of these statements is true.

Answer: (a)
MC   Easy

31.38: In the fission reaction $^{235}U + {}^{1}n \rightarrow {}^{141}Ba + {}^{92}Kr + $ neutrons, the number of neutrons produced is
a) zero.   b) 1.   c) 2.   d) 3.

Answer: (d)
MC   Moderate

31.39: In a nuclear power reactor of the type used to generate electricity, a neutron bombards a uranium nucleus, causing it to split into two large pieces (fission fragments) plus two or three neutrons. Energy is released in this process in the form of electromagnetic radiation and kinetic energy. Which of the following is an accurate statement concerning what happens in such a fission process?
   a) The electrical energy generated comes from the kinetic energy of the incident neutrons.
   b) The electrical energy generated comes from chemical energy stored in the electron bonds of the uranium atom.
   c) Electrical energy is generated because the fission fragments are electrically charged, whereas the uranium was electrically neutral.
   d) An intermediate step involves the fusion of protons to form helium nuclei, and energy is released in this process.
   e) The total mass of the particles after fission is less than the total mass of the particles (uranium nucleus plus one neutron) before fission, and this decrease in mass Δm is converted into energy E, where $E = \Delta mc^2$.

Answer: (e)
       MC   Moderate

31.40: What is the principle difference between a hydrogen bomb and a uranium bomb?
   a) A uranium bomb is an atomic bomb, and a hydrogen bomb is a nuclear bomb.
   b) A uranium bomb utilizes a fission reaction whereas a hydrogen bomb utilizes a fusion reaction.
   c) Both work on the same principle, but the hydrogen bomb has a higher yield.
   d) A hydrogen bomb converts mass into energy, whereas a uranium bomb does not.
   e) One results in radioactive fallout, and the other does not.

Answer: (b)
       MC   Easy

31.41: One sometimes hears reference to a "20-kiloton" bomb. What does this mean?
   a) It means 20,000 tons of nuclear explosive is used in the bomb.
   b) It means that the number of "fissile" nuclei is equal to the number of trinitrotoluene molecules in 20,000 tons of TNT explosive.
   c) It means that the total weight of the bomb (not just the uranium) is 20,000 tons.
   d) It means that the energy released by the bomb is equal to the energy released when 20,000 tons of TNT is exploded.
   e) This refers to the maximum pressure generated by the bomb when it explodes.

Answer: (d)
       MC   Easy

31.42: What is the meaning of the term "critical mass"?
a) This refers to the mass of the "critical" elements in a reactor, i.e., the uranium or plutonium.
b) This refers to the minimum amount of fissionable material required to sustain a chain reaction.
c) This is the amount of mass needed to make a power reactor economically feasible.
d) This is the material which is just on the verge of becoming radioactive.

Answer: (b)
MC   Easy

31.43: The energy radiated by a star, such as the sun, results from
a) beta decay.   b) alpha decay.
c) fission reactions.   d) fusion reactions.

Answer: (d)
MC   Easy

Elementary Particles

32.1: A deuteron has the same charge as a proton but approximately twice the proton mass. Suppose deuterons are being accelerated in a cyclotron in an orbit of radius 75 cm at a frequency of 8 Mhz. What magnetic field is needed?

Answer: 1.07 T
ES   Moderate

32.2: What happens to the cyclotron frequency of a charged particle if its speed doubles?
   a) It triples.   b) It doubles.
   c) It halves.   d) It does not change.

Answer: (d)
MC   Easy

32.3: A cyclotron operates at 10 MHz. What magnetic field is needed to accelerate protons?
   a) 0.12 T   b) 0.25 T   c) 0.66 T   d) 1.12 T

Answer: (c)
MC   Easy

32.4: Is the following decay permissible? If not, state why.
$\Sigma^+ \rightarrow \Lambda^0 + \pi^+$

Answer: No, mass is being created.
ES   Moderate

32.5: Is the following decay allowed? If not, state why.
$\Sigma^+ \rightarrow p + \pi^+$

Answer: Yes.
ES   Moderate

32.6: Is the following decay allowed? If not, state why.
$\pi^0 \rightarrow \mu^+ + \nu_\mu$

Answer: No, charge is not conserved.
ES   Moderate

32.7: Identify the missing particle in the following reaction.
$\pi^- + p \rightarrow \pi^- + \pi^- + \pi^+ + ?$

Answer: p
ES   Moderate

32.8: Identify the missing particle in the following reaction.
$\pi^+ \rightarrow \mu^+ + ?$

Answer: ν_μ
ES   Moderate

32.9: Identify the missing particle in the following reaction.
$\pi^+ + \Lambda^0 \rightarrow \Sigma^+ + ?$

Answer: π^0
ES   Moderate

32.10: What is the quark composition of a neutron?

Answer: udd
ES   Easy

32.11: What is the quark composition of a Λ^0?

Answer: uds
ES   Moderate

32.12: What is the quark composition of a π+ meson?

Answer:
$u\bar{d}$
ES   Moderate

32.13: Particles with integer spin are called mesons.

Answer: False
TF   Moderate

32.14: Particles with half-integer spin are called fermions.

Answer: True
TF   Moderate

32.15: The Pauli exclusion principle does not apply to bosons.

Answer: True
TF   Moderate

32.16: All quarks are fermions.

Answer: True
TF   Moderate

32.17: All hadrons consist of three quarks.

Answer: False
TF Moderate

32.18: Baryon number is conserved in weak interactions.

Answer: False
TF Moderate

32.19: Lepton number is conserved in all interactions.

Answer: True
TF Moderate

32.20: Particles which carry a force are intermediate bosons.

Answer: True
TF Moderate

32.21: The bottom quark has not yet been detected.

Answer: False
TF Moderate

32.22: Strangeness is conserved in all interactions.

Answer: False
TF Moderate

32.23: Z particles are exchanged in weak interactions.

Answer: True
TF Moderate

32.24: The strong force acts on leptons.

Answer: False
TF Moderate

32.25: An anti-quark may have charge $+(2/3)e$.

Answer: False
TF Difficult

32.26: A quark may have charge $-(1/3)e$.

Answer: True
TF Moderate

32.27: Suppose you were to try to create a proton-antiproton pair by annihilation of a very high energy photon. The proton and the anti-proton have the same masses, but opposite charges. What energy photon would be required?

    a) 1.022 MeV   b) 1880 MeV   c) 940 MeV   d) 223 MeV

Answer: (b)
MC   Easy

32.28: A positively charged electron is

    a) called a proton.   b) called a positron.
    c) also called an electron.   d) not found in nature.

Answer: (b)
MC   Easy

32.29: One reason a photon could not create an odd number of electrons and positrons is that such a process would

    a) not conserve charge.
    b) not conserve energy.
    c) require photon energies that are not attainable.
    d) result in the creation of mass.

Answer: (a)
MC   Easy

32.30: When a positronium atom (electron + positron) decays, it emits two photons. What are the energies of the photons?

    a) 0.511 MeV and 1.022 MeV   b) 0.225 MeV and 0.285 MeV
    c) 0.256 MeV and 0.256 MeV   d) 0.511 MeV and 0.511 MeV

Answer: (d)
MC   Easy

32.31: The neutrino has

    a) enormous rest mass, positive charge, and spin quantum number 1/2.
    b) negligible rest mass, negative charge, and spin quantum number 1.
    c) negligible rest mass, no charge, and spin quantum number 1/2.
    d) none of the above.

Answer: (c)
MC   Moderate

32.32: Which of the following is not considered to be one of the four fundamental forces?

    a) The gravity force   b) The meson force
    c) The weak nuclear force   d) The strong nuclear force
    e) The electromagnetic force

Answer: (b)
MC   Moderate

32.33: What effect does an increase in the mass of the virtual exchange particle have on the range of the force it mediates?
a) Decreases it.
b) Increases it.
c) Has no appreciable effect.
d) Decreases charged particle interactions and increases neutral particle interactions.

Answer: (a)
MC   Moderate

32.34: A particle that travels at the speed of light
a) has never been observed.
b) must have zero rest mass.
c) must have a very large rest mass.
d) has infinite energy.

Answer: (b)
MC   Easy

32.35: The exchange particles for the weak force are very massive (about 100 times as massive as a proton). This would lead one to expect that the weak force would
a) act over a very long range.
b) act over a very short range.
c) only act on very massive particles.
d) be transmitted at the speed of light.

Answer: (b)
MC   Moderate

32.36: An electron is an example of
a) a hadron.   b) a meson.   c) a lepton.   d) a baryon.

Answer: (c)
MC   Moderate

32.37: The exchange particle for quarks is called
a) the stickon.   b) the gluon.
c) the epoxyon.   d) the epsilon.

Answer: (b)
MC   Moderate

32.38: An exchange particle for the weak force is the
a) photon.   b) meson.   c) W.   d) graviton.

Answer: (c)
MC   Moderate

32.39: Which of the following is true?
a) All hadrons are baryons or leptons.
b) All hadrons are leptons or mesons.
c) All hadrons are mesons or baryons.
d) All hadrons are nucleons.

Answer: (c)
MC   Moderate

32.40: Nucleons are baryons, not hadrons.

Answer: False
TF   Moderate

32.41: A distinctive feature of quarks is that they
a) have zero rest mass.
b) have zero charge.
c) have fractional electric charge.
d) are always observed singly, since they do not readily interact with other particles.

Answer: (c)
MC   Easy

32.42: A significant step in "unifying" the forces of nature was the discovery that two of the so-called fundamental forces were two parts of a single force. The two forces that were so unified were
a) the electromagnetic force and the weak force.
b) the electromagnetic force and the strong nuclear force.
c) the electromagnetic force and the gravity force.
d) the weak force and the strong nuclear force.

Answer: (a)
MC   Moderate

32.43: The grand unified theory (GUT) is a sought-after model that would unify three of the four forces of nature. Which force is the one it would not include?
a) Gravitational   b) Strong
c) Weak   d) Electromagnetic

Answer: (a)
MC   Moderate

32.44: A quark is
a) a constituent of a nucleon.
b) a constituent of a hadron.
c) an elementary particle.
d) all of the above.

Answer: (d)
MC   Easy

32.45: Mesons consist of quark-antiquark pairs.

Answer: True
        TF   Easy

32.46: The number of quarks in a deuteron ($^2_1$H) is

    a) 1   b) 2   c) 3   d) 4   e) 6

Answer: (e)
        MC   Moderate

32.47:

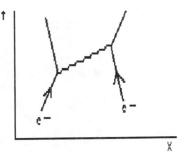

The Feynman diagram shows two
electrons approaching each other,
interacting, then leaving each other.
What particle is being exchanged during the interaction?
    a) Pion   b) Virtual photon
    c) Neutrino   d) W particle

Answer: (b)
        MC   Moderate

32.48:

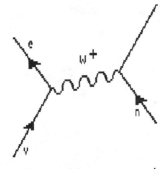

The Feynman diagram shows the weak
interaction of a neutrino (ν) and a
neutron (n), mediated by a W^+ particle.
The interaction produces an electron (e^-).
and another particle. What is the other particle?
    a) A positron (e^+).   b) A pion (π^+).
    c) A proton (p).   d) A quark.
    e) An anti-neutron.

Answer: (c)
        MC   Moderate

32.49: The number of types of quarks (including the "top" quark, which
       has not yet been observed) is
    a) 2.   b) 3.   c) 6.   d) 8.

Answer: (c)
        MC   Moderate

32.50: When a quark emits or absorbs a gluon, the quark changes
      a) its charge.   b) its mass.
      c) its color.   d) into an antiquark.

Answer: (c)
      MC   Moderate

Astrophysics and Cosmology

33.1: The earth's orbit has a mean radius of 1.5×10^8 km. Over a six-month period, the apparent position of a particular star varies by 0.00014° due to parallax. How distant is the star, in both km and ly? (1 ly = 9.46×10^{15} m)

Answer: 1.2×10^{14} km; 13 ly
ES Moderate

33.2: What is the parallax angle for Proxima Centauri, which is earth's nearest star at 4.3 ly?

Answer: 2.1×10^{-4} degrees
ES Moderate

33.3: The sun has has apparent brightness at earth (1.5×10^8 km away) of B. What would be the apparent brightness of the sun at Pluto, which is 6×10^9 km from the sun?
a) 0.25 B   b) 0.063 B   c) 0.025 B   d) 0.00063 B

Answer: (d)
MC Easy

33.4: A Hertzsprung-Russell diagram shows stars on a plot of
a) apparent brightness vs. temperature.
b) luminosity vs. temperature.
c) luminosity vs. distance.
d) magnitude vs. apparent brightness.

Answer: (b)
MC Easy

33.5: Four different main-sequence stars are colored blue, orange, red, and yellow. What is their rank from coolest to hottest?
a) blue, yellow, orange, red   b) orange, blue, yellow, red
c) red, orange, yellow, blue   d) red, yellow, orange, blue

Answer: (c)
MC Easy

33.6: A white dwarf star the mass of the sun is about the size of the earth.

Answer: True
TF Moderate

33.7: The Chandrasekhar limit of stellar mass (below which a star will collapse into a white dwarf) is, if the mass of our sun is M, about

 a) 0.8 M   b) 1.2 M   c) 1.4 M   d) 1.9 M

Answer: (c)
    MC   Moderate

33.8: A star with mass above the Chandrasekhar limit will end up exploding in a supernova.

Answer: True
    TF   Easy

33.9: Supernovas are thought to always result in neutron stars.

Answer: False
    TF   Moderate

33.10: Pulsars are rapidly spinning neutron stars.

Answer: True
    TF   Easy

33.11: It can be shown that the approximate age of the universe is 1/H, where H is the Hubble constant. Taking H = 20 km/s/Mly, estimate the age of the universe, in years.

Answer: 15×10^9 years
    ES   Moderate

33.12: The cosmic background radiation corresponds to a temperature of about

 a) 1.4 K   b) 2.7 K   c) 3.0 K   d) 3.3 K

Answer: (b)
    MC   Moderate

33.13: About 1 μs after the Big Bang, the temperature of the universe was about 10^{13} K. What energy (in eV) does this correspond to?

 a) 1 keV   b) 1 MeV   c) 1 GeV   d) 1 TeV

Answer: (c)
    MC   Easy

33.14: In the standard model history of the universe, the formation of hadrons preceeded the formation of leptons.

Answer: True
    TF   Easy

33.15: If the curvature of the universe is negative, then the universe will eventually stop expanding and collapse in on itself.

Answer: False
TF Easy

33.16: The gravitational red shift is caused by
a) gravitational lensing.
b) Rayleigh scattering in the atmosphere.
c) time dilation.
d) rotating black holes.

Answer: (c)
MC Moderate

33.17: Black holes
a) are holes in space, devoid of matter.
b) are predicted by Einstein's special theory of relativity.
c) are the collapsed remnant of stars.
d) cannot be detected in binary star systems.
e) are a violation of Einstein's general theory of relativity.

Answer: (c)
MC Easy

33.18: The Schwarzchild radius of a black hole is that radial distance from the center of a sphere within which not even light can escape. It was first discovered mathematically by Schwarzchild in 1916 after Einstein published his general relativity theory. It can be calculated from a star's mass M as: $R = 2GM/c^2$. If the mass of star A is twice as much as the mass of star B, the average density of star A, compared to star B will be
a) the same.
b) twice as much.
c) half as much.
d) four times as much.
e) one-fourth as much.

Answer: (d)
MC Moderate

33.19: According to Einstein's general theory of relativity, rotating reference frames cannot be distinguished from gravitational acceleration.

Answer: False
TF Moderate